Revised edition

New Wave Mental Maths G

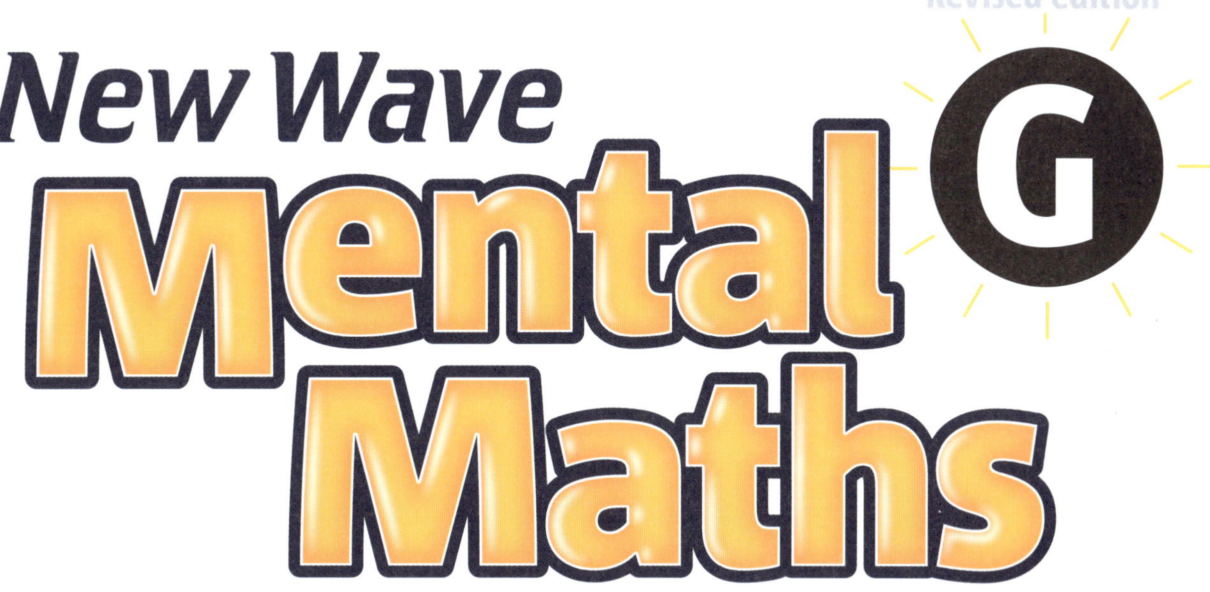

This book belongs to

....

R.I.C. PUBLICATIONS

Eddy Krajcar

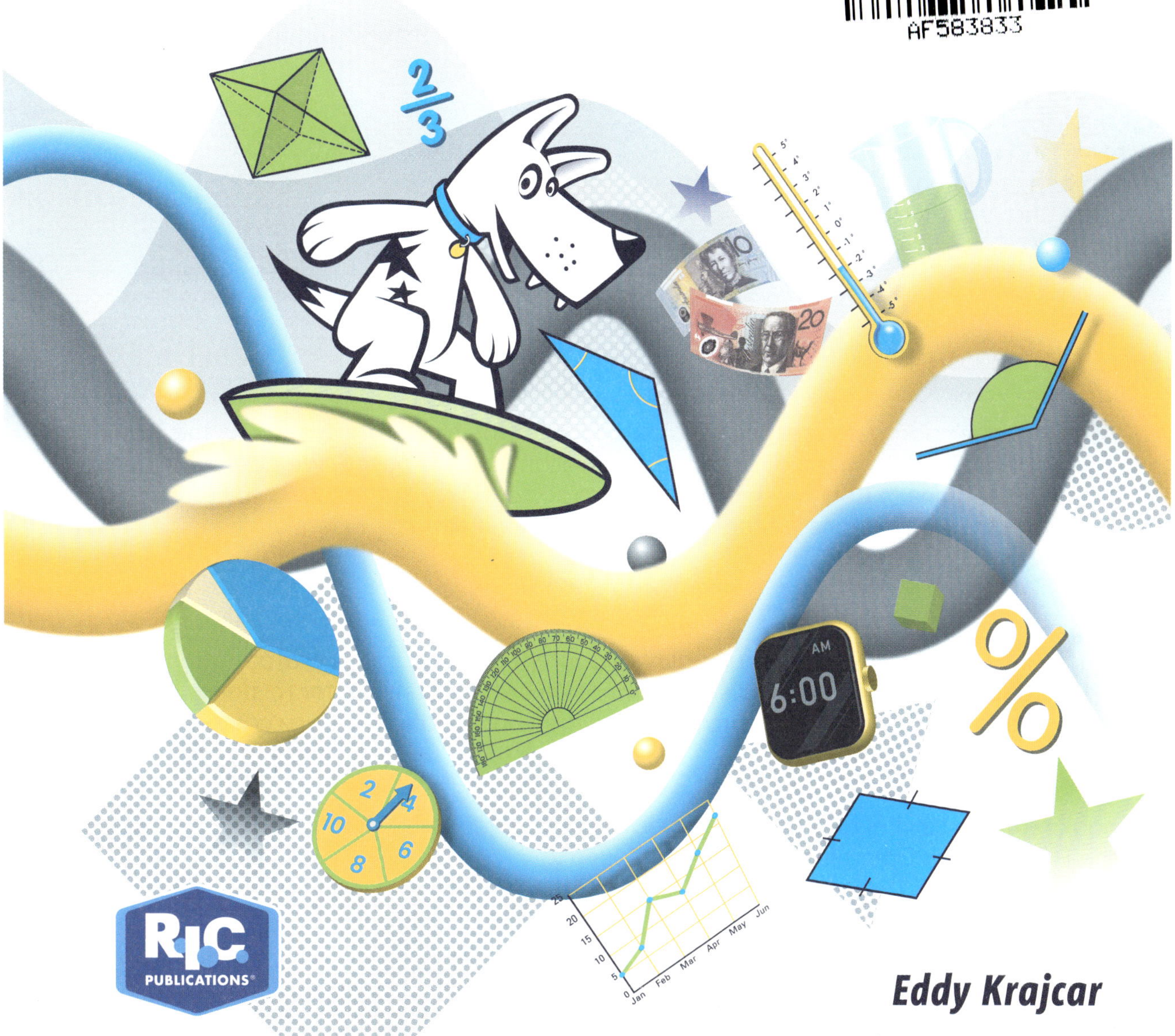

Published by R.I.C. Publications®
PO Box 332, Greenwood
Western Australia 6924
+61 8 9240 9888
www.ricpublications.com.au
mail@ricpublications.com.au

First published 1999
Rewritten and reprinted 2011
Revised and reprinted 2012
Rewritten and reprinted 2024
Reprinted 2025

RIC-8620

ISBN 978-1-923005-54-9

R.I.C. Publications® acknowledges the Wadjak people of the Nyoongar Nation as the Traditional Custodians of the land on which our Western Australian office is based. We acknowledge the Traditional Custodians of Country throughout Australia and pay our respects to Elders past and present. R.I.C. Publications® recognises the role of First Nations Elders as Australia's first educators.

Disclaimer
Every effort has been made to ensure quality of content and accuracy of information; our team at R.I.C. Publications® and Prim-Ed Publishing cannot be held responsible for mistakes or omissions, but we do endeavour to rectify any errors found within our products. Please contact us to provide feedback.

This book features several artworks illustrated and approved for appropriateness and accuracy by Melinda Brown, Spirit Dreaming, Ngunnawal Country.

Foreword

New Wave Mental Maths is a best-selling series written to consolidate students' mathematical understanding through comprehensive and structured daily mental maths practice. The series supports learning for curricula across Australia, and has been revised to meet the proficiencies of the Australian Curriculum Version 9.0, covering Understanding, Fluency, Problem-solving, and Reasoning.

The seven workbooks are designed to ensure students:

- practise mental calculation concepts and skills across all strands
- consolidate their understanding of mathematical topics and vocabulary
- reinforce their learning through self-guided revision of pre-taught concepts.

New to this edition is the 'Weekly Focus', which highlights a key topic covered throughout the week's learning, offering an ideal opportunity for class discussion. Additionally, the updated 'Problem-solving' column provides students the space to apply critical thinking and inquiry skills to in-depth problems.

The 'Friday Review' consolidates the week's learning with colour-coded questions representing the curriculum strands: blue for Number, purple for Algebra, green for Measurement, orange for Space, red for Statistics, and black for Probability. Designed to be flexible, educators may choose to use the 'Friday Review' as a pre- or post-assessment, or to allow students to self-assess and monitor their own progress.

The 'Maths Facts', located at the back of each workbook, support student learning through visual representation of mathematical concepts.

The goal of this series is to encourage students to practise mental calculations every day, supporting their future in a world that is full of maths.

Contents

Week 1

Monday

1. $\frac{1}{4}$ = 0.______ = ______%

2. 8 × 5 = ______, 0.8 × 0.5 = ______

3. If a clock shows 9 o'clock, what is the size of the smaller angle between the two hands?

   45° ☐ 90° ☐ 9° ☐

4. $\frac{4}{5} + \frac{4}{5}$ = ______

5. The opposite numbers on a die always add up to seven. Of which two numbers could the blank face be?

   ______ ______

6. The prime numbers < 10 are:

   ______, ______, ______, ______.

7. What is the lowest common denominator (LCD) for $\frac{1}{2}$ and $\frac{1}{5}$?

   (a) 8 ☐ (b) 10 ☐ (c) 20 ☐ (d) 7 ☐

8. Simplify $\frac{5}{10}$. ______

9. Name this shape.

   ______

10. Is 213 divisible by 3 or 4? ______

11. If rounded, is $3\frac{3}{4}$ closer to 3 or 4? ______

12. Measure the line below. ______mm

13. Write 6 pm in 24-hour time.

    ______

14. What is the area of a square with 4cm sides?

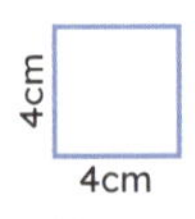

(Not to scale.)

    ______

15. 4 – 0.2 = ______

Tuesday

1. $\frac{3}{4}$ = 0.______ = ______%

2. Draw a reflection.

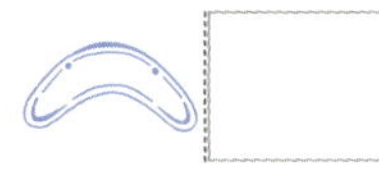

3. Round $2\frac{7}{10}$ to the nearest whole number.

   ______

4. Follow the instructions to correctly write 23, 11, 9, and 55 in the correct position.

   (a) The lower prime number has no number to its left.

   (b) The two composite numbers are between the two prime numbers.

   (c) The multiple of five is immediately left of a prime.

   ______, ______, ______, ______

5. Halve 500. ______

6. If 16 = $(a + 8)$, then a = ______

7. (5 × 5) + (2 × 10) = ______

8. Draw the square's diagonals.

9. The lowest common denominator (LCD) for $\frac{1}{4}$ and $\frac{3}{10}$ is:

   (a) 10. ☐ (b) 15. ☐ (c) 20. ☐ (d) 24. ☐

10. The place value of the eight in 82 411 is

    ______.

11. Olivia puts eight roses in each of six vases. She has two roses left over. How many roses did Olivia begin with?

    ______

12. $\frac{6}{10} + \frac{4}{100}$ = 0.______

13. $5.50 + $6.50 = ______

14. Which two towns are 160km apart?

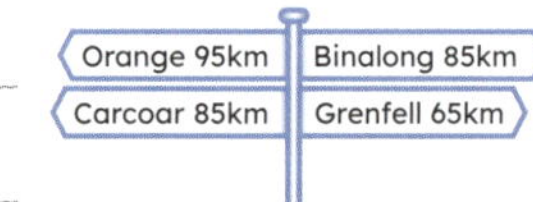

    ______

    ______

15. $5\overline{)105}$ = ______

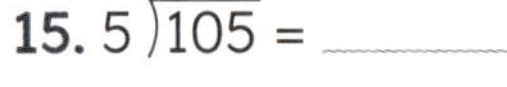

Wednesday

1. $\frac{3}{5}$ = 0.________ = ________%

2. Write C midway of $\overline{AB}$. The length of line $\overline{AC}$ is ________mm.

A B

3. 4, 9, 14, ________, 24, ________

4. Write one million and ten as a numeral.

____

5. 6 – 0.3 = ________

6. Write 9:00 pm in 24-hour time.

____

7. Ren puts six tennis balls in each of seven baskets. There are three balls left over. How many balls did Ren start with?

____

8. What comes next in the pattern?

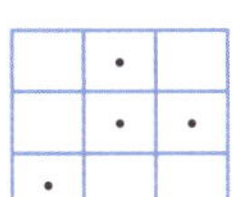 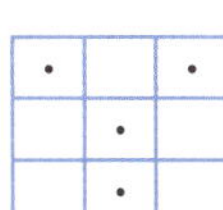 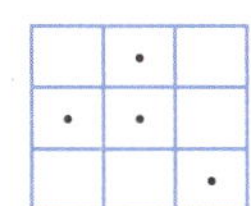 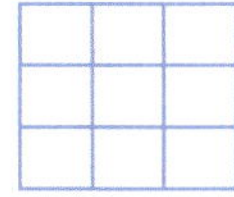

9. Simplify $\frac{6}{9}$. ________

10. If a clock shows 3 o'clock, what is the size of the smaller angle between the hands?

45° ☐ 90° ☐ 180° ☐

11. 20% of $100 = ________

12. 8 × 12 = 4 × ________ = 2 × ________ = 1 × ________

13. Halve 90. ________

14. If a jar contains four chocolate chip biscuits, six Anzac biscuits, and 10 coconut biscuits, what is the probability of randomly picking a chocolate chip biscuit?

________ (Fraction and decimal.)

15. The value of the eight in 85 633 is

________.

Thursday

Week 1

1. $\frac{7}{10}$ = 0.________ = ________%

2. $\frac{3}{4} + \frac{3}{4}$ = ________

3. Draw the pentagon's diagonals.

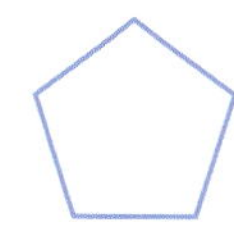

4. Share $80.00 among 16 people.

$________ each

5. (3 + 5) × (3 + 4) = ________

6. Round $\frac{33}{7}$ to the nearest whole number.

____

7. What is the mode of 20, 22, 21, 20, 23, and 25?

____

8. A square has four ________° angles.

9. 8 ÷ 2 = 4, 80 ÷ 20 = ________

10. 8000 + 9000 = ________

11. A pentagon has ________ sides.

12. If you can ride your bike 3km in 10 minutes, how many kilometres can you go in one hour?

________km

13. 12 × 10 = 5 × ________

14. Write ten million one hundred as a numeral.

____

15. Which two towns are 144km apart?

____

____

Elleker 67km
Bow Bridge 57km
Albany 57km
Walpole 77km

Week 1

Problem-solving

Katie baked 200 cupcakes for an engagement party. $\frac{1}{5}$ of the cupcakes were decorated with golden rings, 0.25 had cupids, 12.5% had roses, 0.3 had hearts, and $\frac{1}{8}$ had kissing lips.

Write, as a fraction, decimal, and percentage, how many of each cupcake design there were.

Read the question again. **Think** about the information. Underline the important words.

Tick the strategy you will use to work out the answer:

- estimate and check ☐
- look for patterns ☐
- draw a diagram or picture ☐
- construct a table or graph ☐
- use materials ☐
- use a formula ☐
- something else. ☐

Solve it:

Reflect on the question and answer.

Check it. Circle another strategy on the list to work it out.

Show it:

Friday Review

1. $\frac{4}{5}$ = 0.______
   = ______%
2. Following the clues, label each answer line as either 35, 36, or 45.
   (a) The multiple of 5 and 7 is on the right.
   (b) The multiple of 9 and 5 is on the left.
   ______ ______ ______
3. $5\overline{)135}$ = 135 ÷ 5
   = (100 ÷ 5) + (35 ÷ 5)
   = ______
4. If 20 = (5 + a), then
   a = ______.
5. $5\overline{)205}$ = ______
6. 6 × 8 = 12 × ______
7. 3 × 6 – 2 = ______
8. In Wednesday question 14, what is the probability of randomly picking a coconut biscuit?
   ______
9. $\frac{3}{4} + \frac{3}{4}$ = ______
10. If you ride your bike 4km in 10 minutes, how many kilometres can you travel in one hour?
    ______
11. What is the range of 20, 22, 21, 20, 23, and 25?
    ______
12. Name this shape.
    ______
    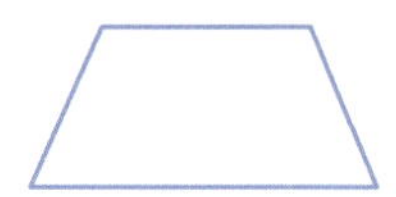
13. (4 + 8) × (3 + 3)
    = ______
14. $\overline{AB}$ is ______mm long.
    A —————— B
15. $\frac{1}{3}$ = 0.______
    = ______%
16. Alana has six roses in each of eight vases. There are five left over. How many roses altogether?
    ______
17. Draw the pentagon's diagonals. How many are there?
    ______
    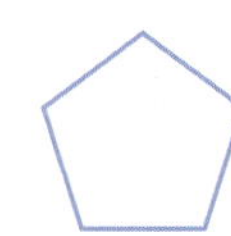
18. Which two towns are 30 kilometres apart?
    ______
    ______

Flint 123km
Copley 193km
York 223km

N A M Sp St P

Monday

1. Which type of triangle?

   (a) equilateral ☐

   (b) isosceles ☐

   (c) scalene ☐

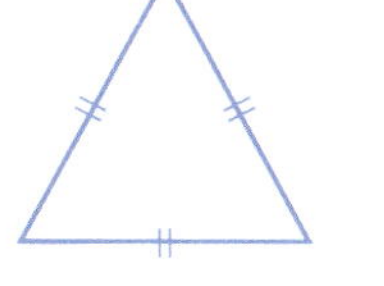

2. Alex exchanged A$15 for EUR€10. How many A$ would Alex exchange for EUR€100?

   ____

3. Name this 3D object.

   ____

4. 6500 + 9500 = ____________

5. What is the place value of the 9 in 962 000?

   ____

6. Continue the pattern.

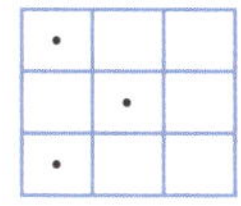 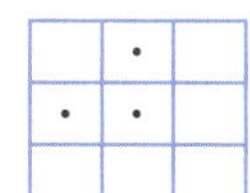 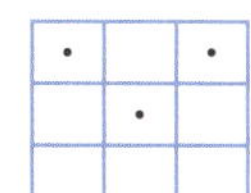 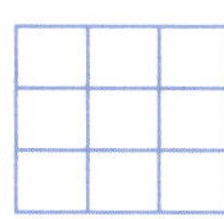

7. If a clock shows 6 o'clock, what is the size of the angle created by the hands?

   90° ☐ 45° ☐ 60° ☐ 180° ☐

8. 130 ÷ 5 = (100 ÷ 5) + (30 ÷ 5) = ____________

9. The prime numbers $> 10 < 20$ are:

   ________, ________, ________, ________.

10. 30% of $50.00 is ________.

11. 6 ÷ 3 = 2, 60 ÷ 30 = ____________

12. How many 20c coins make up $6.80?

    ____

13. This regular pentagon has 7cm-long sides. What is its perimeter?

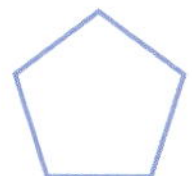

    ____

14. $4.04 > 4.20$ true ☐ false ☐

15. If a bus leaves its depot at 9:05 am and arrives at its first destination at 9:17 am, what is the travelling time?

    ____

Tuesday

1. Which type of triangle?

   (a) equilateral ☐

   (b) isosceles ☐

   (c) scalene ☐

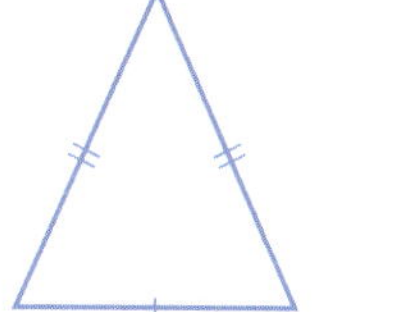

2. 85 + 135 = ____________

3. 145 ÷ 5 = (100 ÷ 5) + (________ ÷ 5)

   = 

4. Draw as a $\frac{3}{4}$ turn clockwise.

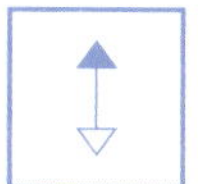 

5. $\frac{3}{4} + \frac{2}{4} + \frac{3}{4} =$ ____________

6. $6^2 =$ ________ × ________ = ________

7. 2017 – 100 = ____________

8. 3 × 10 = 6 × ____________

9. What would be the area of a 50m by 60m rectangle-shaped field?

   ____

10. Name this 3D object.

    ____

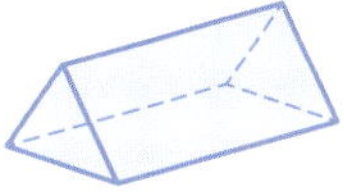

11. 8.75, ________, 9.25, 9.50, ________, 10.00

12. 23.5 ÷ 10 = ____________

13. 150, ________, 450, ________, 750, 900

14. Milky Pool is 150km from Blue Pool and 40km from Green Pool. Write the correct distance for the sign.

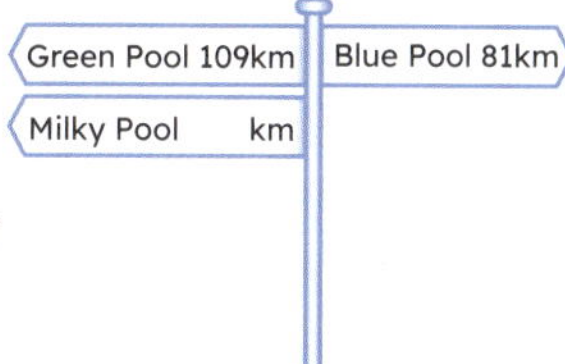

    ____

15. Round 8.6 to the nearest whole number.

Wednesday

Week 2

1. What is the sum of the interior angles of a regular pentagon?

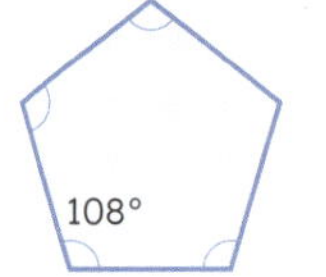

_______ °

2. Which decimal is between $\frac{1}{2}$ and $\frac{3}{4}$?

(a) 0.4 ☐ (b) 0.5 ☐ (c) 0.6 ☐ (d) 0.07 ☐

3. $\frac{12}{10} + \frac{8}{100} =$ _____._____

4. 6.05 < 6.60 true ☐ false ☐

5. Continue the pattern.

A		A
		B
		C

	A	
A		
		BC

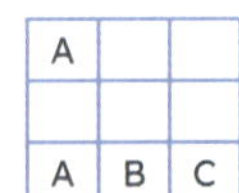

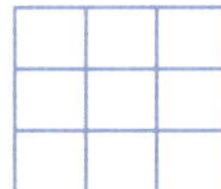

6. What is the value of the four in 417 200?

____

7. Measure the length of line $\overline{AB}$. _______ mm

8. 40% of $80 = _______

9. Draw the pentagon's diagonals. How many are there?

____

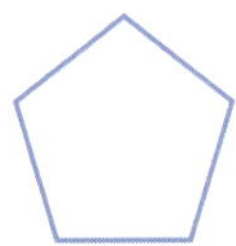

10. $6\overline{)126}$ = (120 ÷ 6) + (6 ÷ 6) = _______

11. 25, 50, _______, 100, _______

12. What is the range of 5, 9, 3, 13, 4, 12, and 5?

____

13. What is the LCD for $\frac{3}{8}$ and $\frac{1}{2}$? _______

14. This regular hexagon has 9cm-long sides. What is its perimeter?

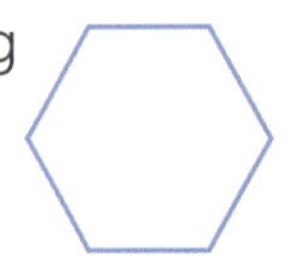

____

15. How many 20c coins make up $4.40?

____

Thursday

1. A regular octagon has _____ sides and an internal angle size of:

(a) 25°. ☐ (b) 90°. ☐

(c) 135°. ☐ (d) 230°. ☐

2. 60% of $50 is _______.

3. The LCD for $\frac{3}{4}$ and $\frac{2}{3}$ is _______.

4. 147 ÷ 7 = (140 ÷ 7) + (7 ÷ 7) = _______

5. $\frac{8}{10} - \frac{3}{10} =$ _______

6. If a bus drives from 8:02 am until 8:13 am, what is the travelling time?

____

7. (a) 16 − 7 = _______ (b) 1.6 − 0.7 = _______

8. 12 ÷ 4 = 24 ÷ _______

9. Draw a $\frac{1}{2}$ turn.

10. If you ride your bike $2\frac{1}{2}$ kilometres in 10 minutes, how many kilometres can you travel in one hour?

____

11. What is the chance of randomly picking a blue pen if your pencil case holds four orange and six blue pens?

_______ in _______

12. Follow the instructions to correctly write 6, 24, 48, and 56 onto each answer line.

(a) The number on the far left has the most factors.

(b) The number on the far right has the least factors.

(c) 56 is to the right of 48.

_______ _______ _______ _______

13. Does 5.27 round to 5.2 or 5.3? _______

14. 35, 70, 105, 140, _______

15. A decagon has _______ sides.

Problem-solving

Consider the similarities and differences among the properties of equilateral, isosceles, scalene, right-angled, acute, and obtuse triangles.

Can each of the above triangle types be constructed using a mix of 3cm, 4cm, and 5cm paper strips?

Read the question again. **Think** about the information. Underline the important words.

Tick the strategy you will use to work out the answer:

- estimate and check ☐
- look for patterns ☐
- draw a diagram or picture ☐
- construct a table or graph ☐
- use materials ☐
- use a formula ☐
- something else. ☐

Solve it:

Reflect on the question and answer.

Check it. Circle another strategy on the list to work it out.

Show it:

Friday Review

Week 2

1 Which type of triangle?
(a) equilateral ☐
(b) isosceles ☐
(c) scalene ☐

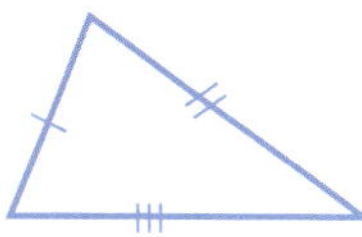

2 $21 \div 6 =$ ______

3 $4 \times 7 = 14 \times$ ______

4 $7^2 =$ ______ $\times$ ______
$=$ ______

5 A bus leaves its depot at 6:04 am and arrives at its first destination at 6:25 am. What is the travel time?

______

6 0.04, 0.08, ______, 0.16

7 What is the sum of the interior angles of a regular pentagon?

______ °

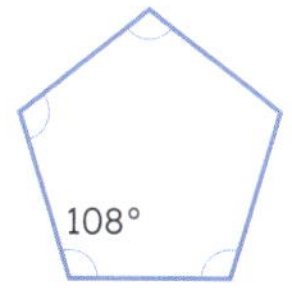

8 40% of $60.00
$=$ ______

9 Draw a $\frac{3}{4}$ turn anticlockwise.

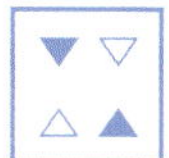

10 Round 7.6 to the nearest whole number.

______

11 The perimeter of a regular pentagon with 7cm sides is ______.

12 What is the median of 5, 9, 3, 13, 4, 12, and 5?

______

13 The value of the four in 411 200 is ______.

14 $9750 + 3250 =$ ______

15 Simplify $\frac{12}{15}$. ______

16 $7\overline{)154} = (140 \div 7) +$
$(14 \div 7) =$ ______

17 $7 - 0.3 =$ ______

18 If a lolly jar contains, four red, 10 yellow, and six orange sweets, what chance have you of picking a red sweet?

______ in ______

M Sp St

978-1-923005-54-9

Week 3

Monday

1. What is the ratio of oranges to apples in a fruit bowl if there are six oranges and three apples?

 ____

2. $\frac{4}{5} + \frac{3}{5} + \frac{4}{5} =$ _______ $=$ _______

3. Tick the approximate size of this angle.

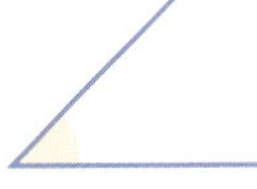

 (a) 20° ☐ (b) 45° ☐ (c) 80° ☐

4. Is 494 evenly divisible by 5? _______

5. If $18 + a = 3 \times 9$, then $a =$ _______.

6. $9 - \frac{1}{5} =$ _______

7. $9^2 =$ _______ × _______ = _______

8. 8 = 48 ÷ 6 = 96 ÷ _______

9. Continue the pattern.

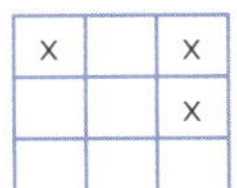 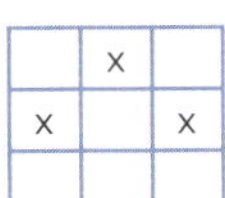 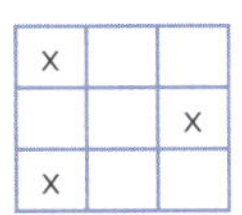 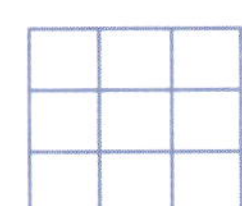

10. 0.03, 0.08, _______, 0.18

11. What is the likely length of the car?

 $5\frac{1}{2}$m ☐ 5.5cm ☐ 550mm ☐

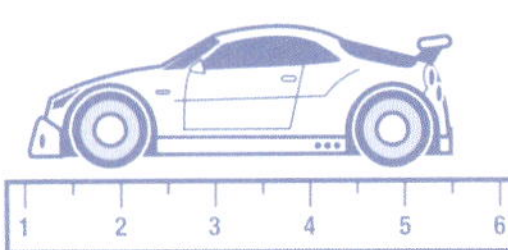

12. Double 8.6. _______

13. If you rode your bike $3\frac{1}{2}$km in a quarter of an hour, how many kilometres would you ride in two hours?

 _______km

14. Round 6.09 to the nearest tenth. _______

15. (a) 47 – 9 = _______ (b) 4.7 – 0.9 = _______

Tuesday

1. What is the ratio of oranges to apples in a fruit bowl if there are eight oranges and two apples?

 ____

2. Draw the hexagon's diagonals. How many are there?

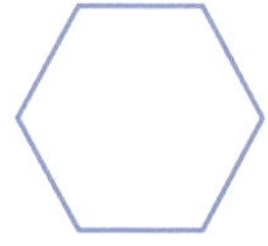

 ____

3. $1\frac{1}{5} - \frac{4}{5} =$ _______

4. (6 × 5) – (2 × 9) = _______

5. 5 – 0.05 = _______

6. $7\overline{)294}$ = (280 ÷ 7) + (14 ÷ 7) = _______

7. What is the ratio of adults to children if a dance school has 20 adults and 10 children?

 ____

8. Draw the reflection.

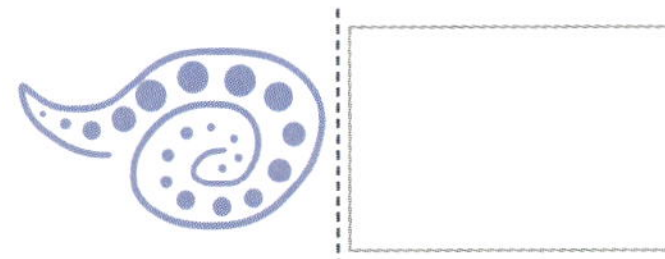

9. What would be the total cost for a family of five (two adults and three children)?

 Adults: $12 Children: $8 $_______

10. Halve 12.9. _______

11. 3, 9, 16, 24, _______

12. If a clock shows 6 o'clock, what is the angle between the hands?

 _______°

13. Round $3\frac{4}{9}$. _______

14. 5 × 7 = (a) 10 × 3.5 ☐
 (b) 10 × 7 ☐
 (c) 10 × 14 ☐
 (d) 10 × 2.5 ☐

15. A store has a 10% discount on all items. What would you pay for a solar-powered pencil sharpener, priced at $50?

 $_______

Wednesday

1. What is the ratio of bananas to pears in a fruit bowl if there are eight bananas and four pears?

   ____

2. If a bucket contains two red balls, five green balls, and 13 yellow balls, what chance have you of choosing a red ball?

   ________ in ________

3. If $200 - b = 110$, then $b =$ ________.

4. Simplify $\frac{15}{18}$. ________

5. Double $\frac{3}{4}$. ________

6. Circle the size of this angle.

   (a) 90° (b) 10° (c) 120° (d) 160°

7. Mr Money paid $100 for his new wallet. The retailer made 30% profit, or

   $________.

8. What is the place value of 5 in 5 111 200?

   ____

9. 10 × 1000 = ________

10. A 20% discount on a $100 pair of shoes means you should pay

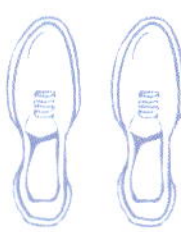

   $________.

11. Which score was only achieved by one student?

   ____

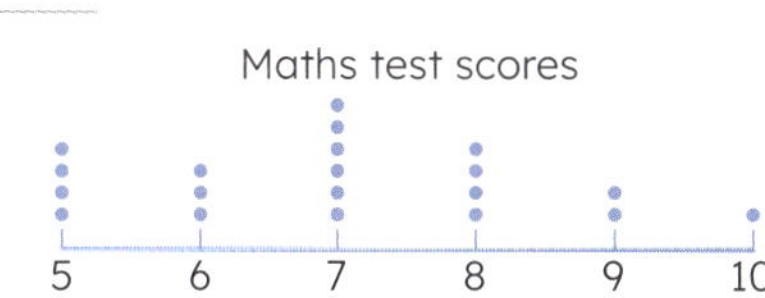

12. 29.5 ÷ 10 = ________

13. Which is the time zone for Queensland, Tasmania, the Australian Capital Territory, and New South Wales (excluding Broken Hill)?

   (a) AEST ☐ (b) ACST ☐ (c) AWST ☐

14. 1L = ________mL

15. Find the LCD for $\frac{1}{3}$ and $\frac{3}{5}$. ________

Thursday

Week 3

1. What is the ratio of bananas to pears in a fruit bowl if there are nine bananas and three pears?

   ____

2. 1000 = 10 × ________

3. 1m = ________cm

4. 6 + 7 = ________, 0.6 + 0.7 = ________

5. 3 – 0.05 = ________

6. Double 1.85. ________

7. Name this 3D object.

   ____

8. Which is not a net of a cube? (Circle your answer.)

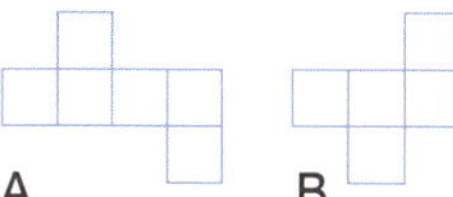
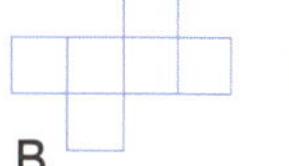
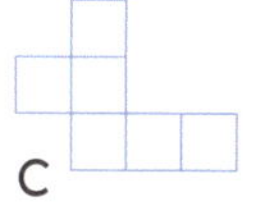
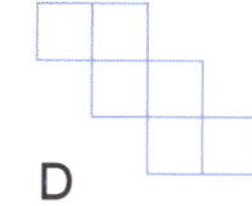

   A B C D

9. (6 × 7) – (4 × 5) = ________

10. $20.00 – $2.25 = $________

11. Angle = ________ °

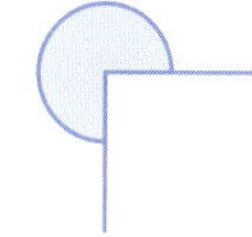

12. How would you write midnight in 24-hour time?

   ____

13. Mrs Money paid $1000 for her new handbag. The retailer's 20% profit on the sale was

   $________.

14. 41.2 ÷ 10 = ________

15. Which expressions are true?

   5 × 9 = (a) 10 × 4.5 ☐ (b) 10 × 18 ☐

   (c) 2.5 × 18 ☐ (d) 5 × 18 ☐

Problem-solving

Dad paid his children, Bill and Ben, a set amount of money per household chore. Each week, a total reward of $120 was available if all the chores were completed. This week, all the chores were completed! The payment was split between Bill and Ben in a ratio of 1:7 respectively.

What fraction of the total reward did Bill get and how much did Bill get paid?

Read the question again. **Think** about the information. Underline the important words.

Tick the strategy you will use to work out the answer:

- estimate and check ☐
- look for patterns ☐
- draw a diagram or picture ☐
- construct a table or graph ☐
- use materials ☐
- use a formula ☐
- something else. ☐

Solve it:

Reflect on the question and answer.

Check it. Circle another strategy on the list to work it out.

Show it:

Friday Review

1. What is the ratio of apples to pears in a fruit bowl if there are six apples and two pears?
   ______

2. $\frac{3}{5} + \frac{4}{5} =$ ______

3. If $15 + a = 2 \times 9$, then
   $a =$ ______

4. A store has a 10% discount. What should you pay for an item originally priced at $80.00?
   $______

5. 10 000 =
   ______ × ______

6. Halve 25. ______

7. (7 × 6) – (2 × 11)
   = ______

8. What is the size of this angle?
   ______°

9. Draw the pentagon's diagonals. How many are there?
   ______

10. Write one million, one thousand, and eleven as a numeral.
    ______

11. 29.5 ÷ 10 = ______

12. If a bucket contains four red balls, seven green balls, and nine yellow balls, what chance have you of choosing a red ball?
    ______

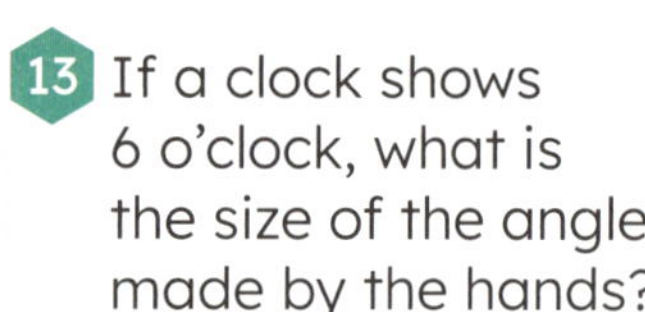

13. If a clock shows 6 o'clock, what is the size of the angle made by the hands?
    ______°

14. What is the LCD for $\frac{2}{3}$ and $\frac{1}{5}$?
    ______

15. Draw the reflection.

16. Round 7.07 to the nearest tenth.
    ______

17. If you can ride your bike $2\frac{1}{2}$km in a quarter of an hour, how far will you ride in two hours?
    ______km

18. Which score was achieved by the most students?
    ______

Maths test scores

5 6 7 8 9 10

N A M Sp St P

Monday

1. Round π (3.14) to the nearest whole. ______

2. Simplify $\frac{20}{25}$. ______

3. 10^3 = ______ × ______ × ______ = ______

4. 600 + 900 = ______

5. 29.3 ÷ 10 = ______

6. 10 – 0.2 = ______

7. What number of diagonals does this heptagon have?

  2 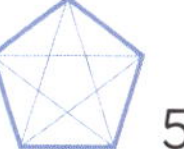 5 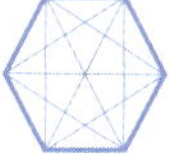 9 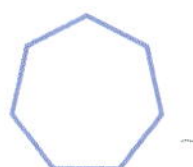 ______

8. 98 + 3 + 5 = ______

9. Round 7.89 to the nearest tenth. ______

10. Draw an arrow to show the concave face.

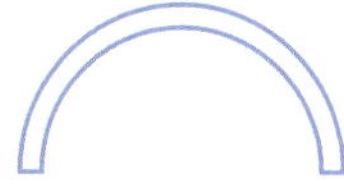

11. ✓ or ✗

5 × 11 = (a) 10 × 11 ☐ (b) 10 × 5.5 ☐

(c) 2.5 × 11 ☐ (d) 2.5 × 22 ☐

12. Is 153 divisible by 9? yes ☐ no ☐

13. 1.2, ______, 0.8, 0.6

14. 77 + 97 = ______

15. A 500g jar of blueberry jam has 40% fruit. How many grams of blueberries are in the jar?

______

Tuesday

1. What is the diameter of a circle if its radius is 6cm?

______

2. Shade 1.25L.

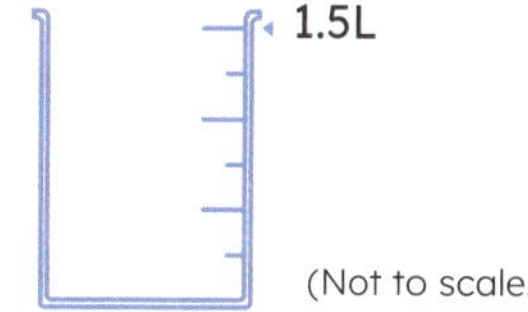

3. $1\frac{3}{4} + 2\frac{3}{4}$ = ______

4. 80 × 500 = ______

5. Which ball has the greatest mass?

(a) 75g ☐ (b) 112g ☐ (c) 1.1kg ☐

6. Write one point one million as a numeral.

______

7. $20.00 – $5.15 = ______

8. 10^4 = ______ × ______ × ______ × ______

= ______

9. 7.70 > 7.09 true ☐ false ☐

10. Measure line $\overline{AB}$ in mm.

______mm

11. 101 – 8 = ______

12. Which type of triangle?

(a) equilateral ☐

(b) isosceles ☐

(c) scalene ☐

(d) scaletrix ☐

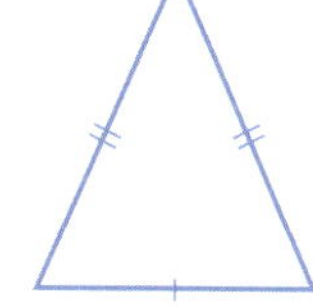

13. Mrs Money paid $600 for her new handbag. The retailer's 25% profit on the sale was

$______.

14. What is the ratio of adults to children if there are six adults and 12 children in a karate class?

______

15. Halve 10.5. ______

Week 4

Wednesday

1. What is the radius of a circle if its diameter is 3m?

   ____

2. Complete the pattern.

A		
C		B

	A	
C		
	B	

	C	
B		A

3. 1m = ____________ mm
4. Is 252 divisible by 9? ____________
5. Round 8.96 to the nearest tenth ____________.
6. $8\overline{)504}$ = ____________
7. 1, 3, 9, 27, ____________, ____________, 729
8. If $a + 25 = 100$, then a = ____________
9. $4\frac{1}{5} - \frac{3}{5}$ = ____________
10. 16 500 + 9500 = ____________
11. 12 × 11 = ____________, 120 × 11 = ____________
12. Shade 1300mL.

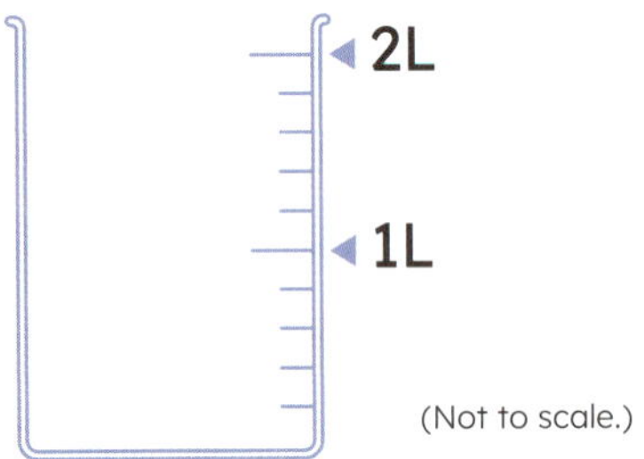

(Not to scale.)

13. 10.6 ÷ 10 = ____________
14. Double $\frac{1}{2}$. ____________
15. Recording collected data can be achieved using a computer program.

    true ☐ false ☐

Thursday

1. What is the radius of a circle if its diameter is 4.6cm?

   ____

2. $3 \times 10^2 = 3 \times$ ____________ = ____________
3. If $b - 45 = 55$, then b = ____________.
4. The LCD for $\frac{3}{4}$ and $\frac{5}{6}$ is ____________.
5. How many 50c coins make up $20.00?

   ____

6. Share $50.00 equally among four friends.

   $____________

7. A decimal located between $\frac{3}{4}$ and 1 is:

   (a) 0.75. ☐ (b) 0.9. ☐ (c) 0.65. ☐

8. Penelope, the perfume collector, has 8 × 50mL and 6 × 25mL of perfume. How many mL total?

   ____

9. If to cut a pizza into halves is one cut and into quarters two cuts, then to cut into eighths

   is ____________ cuts.

10. $6\frac{2}{5} - \frac{4}{5}$ = ____________
11. Write $\frac{1}{2}$ as a decimal. ____________
12. On the number line, write A at 1.5, B at 0.5, and C at –1.5.

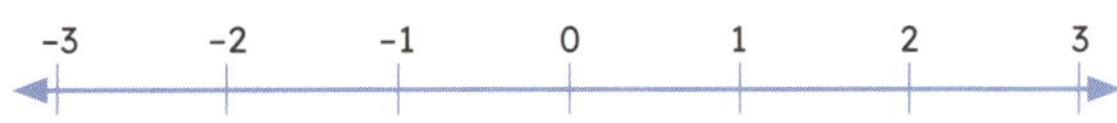

13. Double $\frac{1}{4}$. ____________
14. If a $100 item is reduced by 25%, what is the new price?

    ____

15. 2.97, 2.98, 2.99, ____________

Problem-solving

First Nations Australians use symbols to represent common natural features, places, and objects. Look at this symbol that is often used to represent a camp site in Aboriginal artworks.

What is the approximate circumference of the inner and outer circles?

Read the question again. **Think** about the information. Underline the important words.

Tick the strategy you will use to work out the answer:

- estimate and check ☐
- look for patterns ☐
- draw a diagram or picture ☐
- construct a table or graph ☐
- use materials ☐
- use a formula ☐
- something else. ☐

Solve it:

Reflect on the question and answer.

Check it. Circle another strategy on the list to work it out.

Show it:

Friday Review

1. What is the diameter of a circle if its radius is 8cm? ______

2. Tick the decimal located between $\frac{1}{4}$ and $\frac{1}{2}$.
   (a) 0.35 ☐
   (b) 0.25 ☐
   (c) 0.15 ☐

3. $8\overline{)408}$ = ______

4. Draw an arrow to show the convex face.

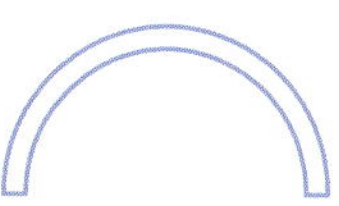

5. 4×10^4 = ______

6. 2.60 > 2.06
   true ☐ false ☐

7. If $a + 55 = 100$, then a = ______.

8. Round 9.98 to the nearest tenth. ______

9. Is 253 divisible by 9?
   yes ☐ no ☐

10. Draw this 2D shape's diagonals. How many are there? ______

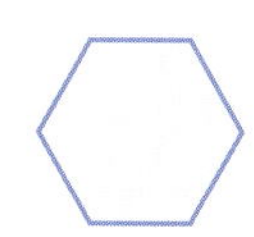

11. $6\frac{1}{5} - \frac{3}{5}$ = ______

12. 78 + 98 = ______

13. Shade 1500mL.

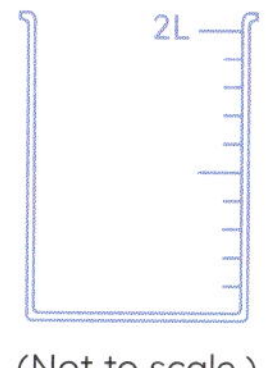

(Not to scale.)

14. A 500g jar of plum jam has 60% fruit. How many grams of fruit in the jar? ______

15. Simplify $\frac{21}{24}$. ______

16. What type of graph could you use to display data you collect about your friends' hobbies?

______

17. Measure line $\overline{XY}$ in mm. ______

X ———— Y

18. A circle has a diameter of 9.8cm. What is its radius? ______

Week 5

Monday

1. Another way of writing 5^2 is:
   (a) 5 × 2. ☐ (b) 5 × 5. ☐
   (c) 52. ☐ (d) 2 lots of 5. ☐
2. One millennium = ______ years
3. How many 20c coins make up $5.00? ______
4. 48 × 5 = 10 × ______
5. 10 – 0.08 = ______
6. Show this witchetty grub after a $\frac{1}{2}$ turn.

7. Halve $\frac{1}{2}$. ______
8. A jar contains 10 blue, four red, and six white beans. What is the chance, as a fraction, of picking a blue bean? ______
9. 7×10^3 = ______
10. $20.00 – $6.35 = ______
11. 81 ÷ ______ = 9
12. $\frac{1}{8}$ = 0.______
13. What is the size of this angle? ______
    It is called a ______ angle.

14. 18 ÷ 2 = ______ ÷ 4
15. 25% of 200 = ______

Tuesday

1. 2^4 = ______ × ______ × ______ × ______ = ______
2. Write the next number after 99 989. ______
3. If a race car's second track time of 49.19 seconds is 0.2 seconds faster than its first, what was the original time? ______

4. Share $120 equally among eight boys. ______ each
5. Share $120 equally among 16 boys. ______ each
6. 1 – 0.04 = ______, 10 – 0.04 = ______
7. Halve $\frac{1}{4}$. ______
8. If you can ride your bike 3km in 12 minutes, how far can you ride in 1 hour? ______
9. Draw a reflection.

10. $7\frac{1}{10} - \frac{8}{10}$ = ______
11. Name this shape. ______

12. What is the ratio of pigs to horses if there are 15 pigs to three horses? ______
13. 25% of 300 is ______.
14. The LCD of $\frac{7}{10}$ and $\frac{3}{4}$ is ______.
15. 8.1, 8.05, 8, ______, 7.90, 7.85

Wednesday

1. 6^2 =

(a) 6 × 2 ☐ (b) 6 × 6 ☐

(c) 2 × 6 ☐ (d) 62 ☐

2. Write the next number after 109 999.

3. 24 × 10 = 5 × ________

4. Continue the pattern.

	x	
		o
x		x

x		x
		o
	x	

	x	
x		o
		x

5. Antonio exchanged A\$50 for EUR€20. How many euros would Antonio receive for A\$75?

6. Simplify $\frac{20}{24}$. ________

7. 2000 – 350 = ________

8. Draw after a $\frac{1}{4}$ turn clockwise.

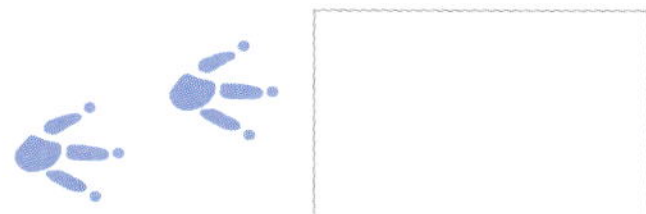

9. One century = ________ years

10. 25% of 500 is ________.

11. A dodecagon has ________ sides.

12. $5\frac{2}{10} - \frac{7}{10}$ = ________

13. A bus timetable shows 8-minute intervals between each stop. If a bus leaves its depot at 9:02 am and has four stops, at what time does it reach its fourth stop?

14. Penelope, the perfume collector, has 12 × 50mL and 7 × 25mL of perfume. How many mL of fragrance is there?

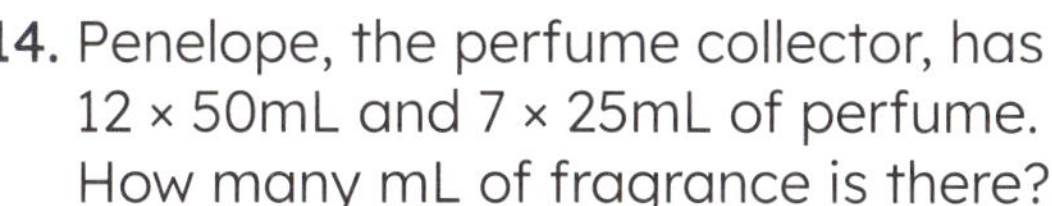

15. How many pens are there altogether?

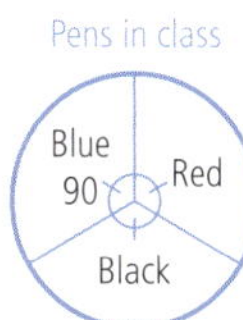

Thursday

Week 5

1. 2^5 = ________

2. If 4 × b = 36, then b = ________.

3. What is the size of this angle? ________

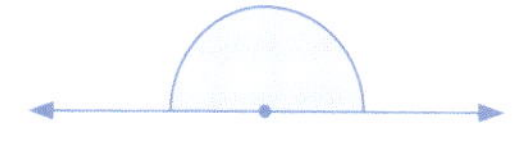

4. 25% of 800 is ________.

5. Is 621 divisible by 9? yes ☐ no ☐

6. 12 × 9 = 3 × ________

7. If a race car's second track time of 48.01 seconds is 0.3 seconds slower than its first, what was the original time?

8. An electrician's invoice shows the following:

Labour cost $80
Light fitting $20
GST @ 10%

(a) What is the GST cost? ________

(b) What is the total cost for the customer?

9. The house sold for \$1.1 million. What was the auctioneer's 10% fee for selling the house?

10. How many 50c coins make up \$15.00?

11. 71.69 ÷ 10 = ________

12. 0.3 + 0.09 + 0.004 = ________

13. Share \$40.00 equally among 16 people.

________ each

14. Draw the pentagon's diagonals. How many are there?

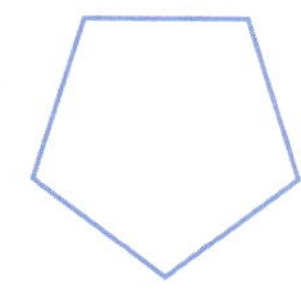

15. A regular octagon has a perimeter of 40cm. What is the length of one side?

Week 5

Problem-solving

A new supersonic airliner has been invented that can travel at 2.4×10^3 km/h.

If the airliner can travel this many kilometres per hour, how many kilometres can it travel in one minute? (Hint: Write the airliner's speed in expanded form first.)

Read the question again. **Think** about the information. Underline the important words.

Tick the strategy you will use to work out the answer:

- estimate and check ☐
- look for patterns ☐
- draw a diagram or picture ☐
- construct a table or graph ☐
- use materials ☐
- use a formula ☐
- something else. ☐

Solve it:

Reflect on the question and answer.

Check it. Circle another strategy on the list to work it out.

Show it:

Friday Review

1. 7^2 =
   (a) 7 × 2 ☐
   (b) 7 × 7 ☐
   (c) 72 ☐
   (d) 14 ☐
2. If $6 \times a = 42$, then a = ________.
3. 25% of 900 = ________
4. How many pens are there altogether?

   ____

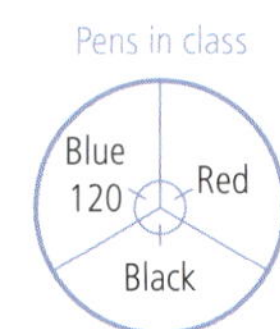

5. Show as a $\frac{1}{4}$ turn anticlockwise.

6. 8^2 =
   (a) 8 × 2 ☐
   (b) 8 + 8 ☐
   (c) 8 × 8 ☐
   (d) 16 ☐
   (e) 82 ☐
7. What is the LCD of $\frac{3}{10}$ and $\frac{3}{4}$?

   ____
8. $\frac{1}{4}$ = ________%
9. $\frac{8}{100} + \frac{7}{10}$ = 0.________
10. The place value of 2 in 0.32 is

11. 16, 32, 64, ________
12. $5\frac{3}{10} - \frac{7}{10}$ = ________
13. 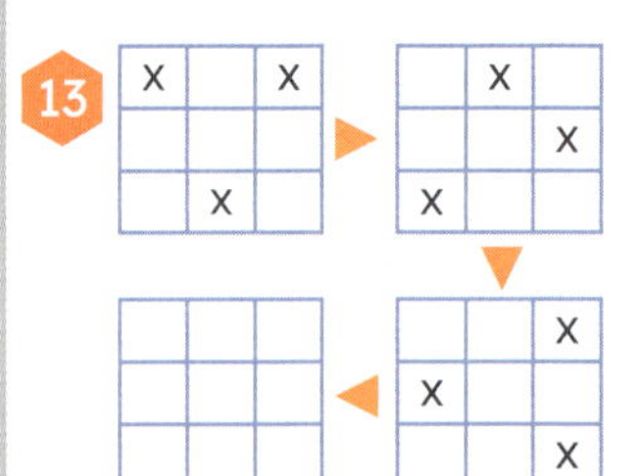
14. A jar contains eight pink, six blue, three green, and three red jelly beans. What is the chance, as a fraction, of picking a blue bean?

    ____
15. 9.1, 9.05, 9, ________, 8.9, 8.85
16. A bus timetable shows 8-minute intervals between each stop. If a bus leaves the depot at 6:04 am and has three stops, at what time will it reach the third stop?

    ____
17. $10^5 \times 5$ =

    ____
18. 4×10^4 = ________

Monday

1. How many B boxes would fit evenly into box A?

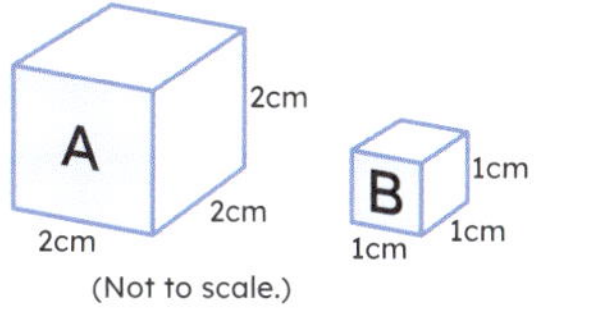

______

2. How many 50c coins make up $20.00?

______

3. Draw and colour a ratio of 3 pink marbles to 1 yellow marble.

4. What is the smallest odd whole number that can be made from 8, 2, 7, and 8?

______

5. What is the area of a 8m-by-6m floor?

______ m^2

6. What is the angle size between the hands of a clock that shows 9 o'clock?

45° ☐ 90° ☐ 9° ☐

7. Halve $\frac{1}{5}$. ______

8. Share $10.00 equally among four people.

______ each

9. 997 + 6 + 9 = ______

10. $20.00 – $4.80 = ______

11. Which two towns are 202km apart?

Forster 136km | Nerong 116km | Wootton 86km | Girvan 126km

______

______

12. $2^5 \times 100$ = ______

13. Order from smallest to largest: 7, 0, –3, –9, 8.

______, ______, ______, ______, ______

14. Write half past one in the morning as 24-hour time.

______

15. Which is true?

(a) –4 > –7 ☐ (b) –4 < –7 ☐

Tuesday

Week 6

1. What is the volume of this block?

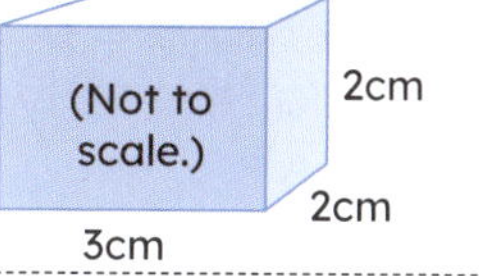

______

2. $0.38 = \frac{\square}{10} + \frac{\square}{100}$

3. Scale: 1cm = 5km. What is the distance from Dubbo to Tubbo?

______

4. What is the range of 50, 20, 80, 50, 30, and 10?

______

5. 90 × 9 = 810, 89 × 9 = ______

6. What is the ratio of ducks to swans if a lake has 10 ducks and 20 swans?

______

7. Round 7.07 to the nearest tenth. ______

8. If $b \times 12 = 108$, then b = ______.

9. 32 ÷ 8 = ______ ÷ 4

10. 8×10^4 = ______

11. A timetable shows 6-minute intervals. If a bus leaves its depot at 6:10 am and has three stops, at what time is the third stop?

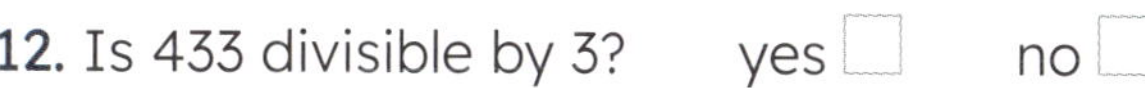

______

12. Is 433 divisible by 3? yes ☐ no ☐

13. How many 20c coins make up $12.00?

______

14. 80 × 50 = ______

15. A hat contains seven red balls and eight blue balls. What is the chance, as a fraction, of picking a blue ball?

______

Week 6

Wednesday

1. How many B boxes would fit evenly into box A?

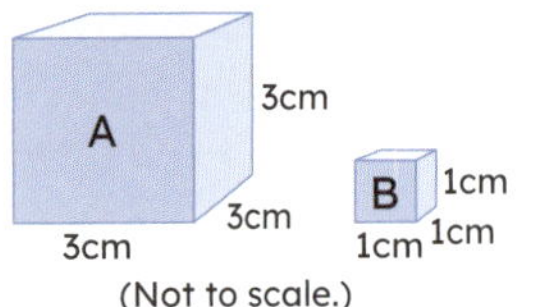

______

2. A classroom has a rectangular perimeter of 30m. One wall is 8m long. What is the floor area?

______

3. 75% of 200 = ______

4. What is the ratio of dogs to cats if a street has 20 dogs and 10 cats?

______

5. Round 29.35 to the nearest whole number.

______

6. Simplify $\frac{21}{24}$. ______

7. Draw a $\frac{1}{4}$ turn clockwise.

8. 3^3 = ______

9. Luke's farm paddock is square and has three fence posts along each side. How many fence posts altogether?

______

10. (a) Write six-hundredths as a decimal.

______

(b) Write six-thousandths as a decimal.

______

11. Name this 3D object.

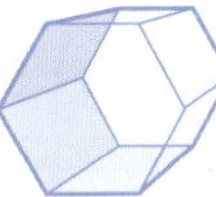

______

12. $8\frac{1}{3} - \frac{2}{3}$ = ______

13. $\frac{1}{4}$ = 0.25, $\frac{1}{8}$ = 0.______

14. (16 – 7) – (12 ÷ 4) = ______

15. Write a at (1,4), b at (2,3), and c at (3,2).

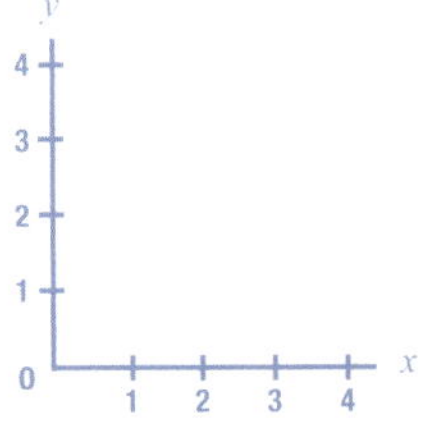

Thursday

1. What is the volume of a box 3m wide by 4m long by 2m high?

______ m^3

2. The prime numbers > 10 but < 20 are:

______, ______, ______, ______.

3. 300, 30, 3, ______

4. Draw a reflection.

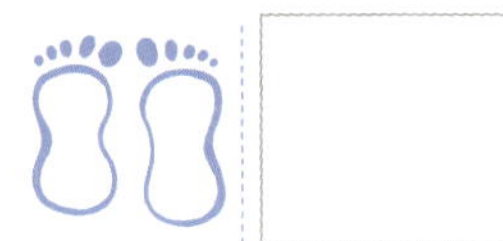

5. If 300 ÷ c = 60, then

c = ______.

6. (a) Which lines are parallel? ______

(b) Which lines are perpendicular? ______

7. Simplify $\frac{30}{36}$. ______

8. (3 × 1000) + (5 + 100) + (2 × 10) + (8 × 1) =

______

9. 26 250 + 4750 = ______

10. Write 1.12 million as a numeral.

______

11. 8.00 > 8.08 true ☐ false ☐

12. What is the number halfway between 18 and 28?

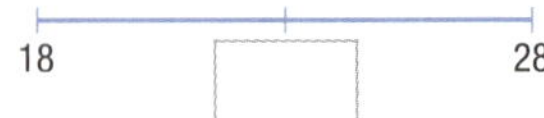

13. 2 – 0.01 = ______, 2 – 0.001 = ______

14. (a) What is the size of angle x? ______

(b) What is the size of angle y? ______

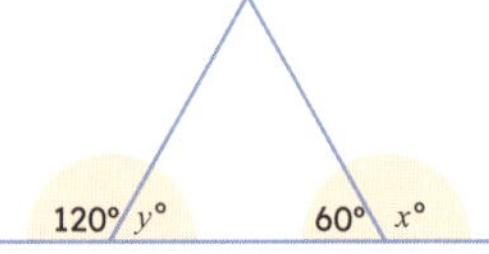

15. How many 20c coins make up $6.80?

______

Problem-solving

Patsy and Bindi built this school model using three cardboard boxes.

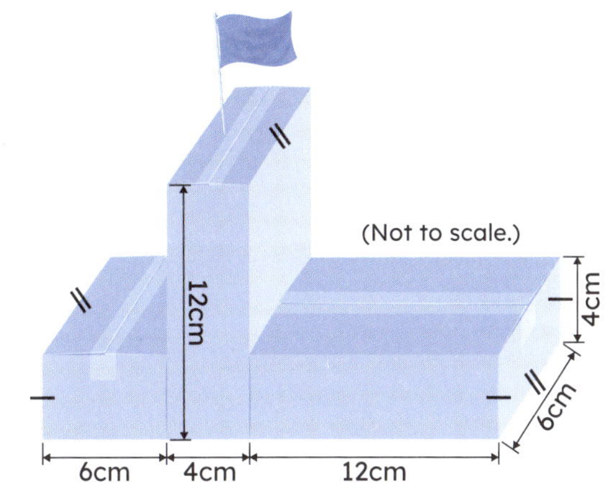

What is the total volume of their school model?

Read the question again. **Think** about the information. Underline the important words.

Tick the strategy you will use to work out the answer:

- estimate and check ☐
- look for patterns ☐
- draw a diagram or picture ☐
- construct a table or graph ☐
- use materials ☐
- use a formula ☐
- something else. ☐

Solve it:

Reflect on the question and answer.

Check it. Circle another strategy on the list to work it out.

Show it:

Friday Review

1. How many B boxes will fit into box A?

____

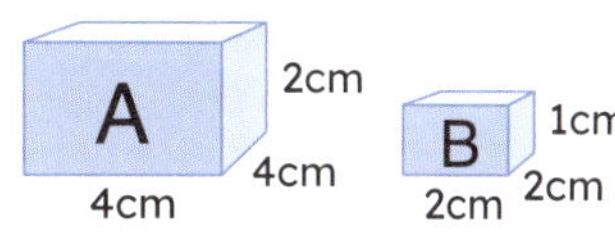

2. What is the ratio of rats to mice if there are 12 rats and 24 mice?

____

3. Draw and colour a ratio of 4 pink and 3 yellow marbles.

4. If $c \times 12 = 132$, then

$c =$ _______.

5. 62.5 ☐ 10 = 6.25

6. $9 - 0.01 =$ _______

7. $100 \div 50 =$ _______

8. $7\overline{)427} =$ _______

9. Write twelve forty in the morning in 24-hour time.

____

10. How many B boxes fit evenly into box A?

____

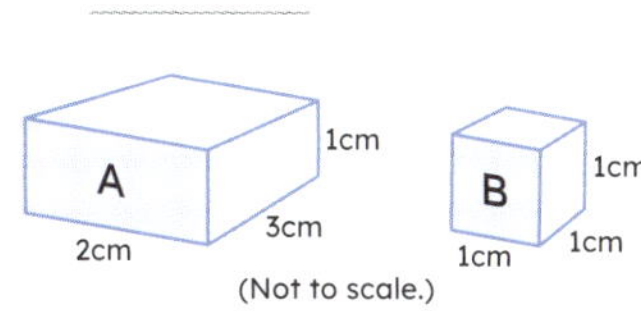

11. How many 20c coins make up $8.40?

____

12. What is the median of 50, 20, 80, 50, 30, and 10?

____

13. $7.07 > 7.70$

true ☐ false ☐

14. Round 32.83 to the nearest tenth.

____

15. $9 \times 105 =$ _______

16. How many edges?

____

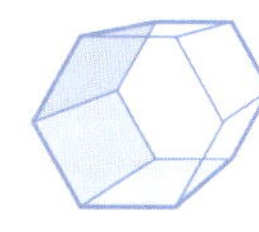

17. $7\frac{6}{8} + 2\frac{4}{8} =$ _______

18. A jar contains three chocolate chip, 12 coconut, and five ginger biscuits. What is the probability of choosing a ginger biscuit?

____

Week 6

Week 7

Monday

1. Add brackets to this number sentence.

   6 × 4 + 3 ÷ 3 = 25

2. Multiply a number by four, double it, subtract six, and the answer is 18. What is the starting number?

   ______

3. Halve $\frac{1}{2}$. ______

4. Draw a $\frac{1}{2}$ turn.

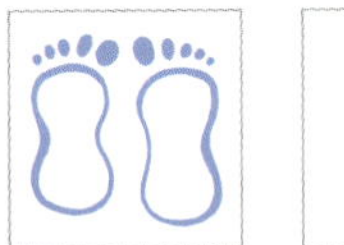

5. 1 – 0.98 = ______

6. 6000 – 2950 = ______

7. Halve 950. ______

8. Your teacher drops a box of 30 eggs. If 10% are broken, how many eggs are okay?

   ______

9. 80 + 70 + 90 = ______

10. How many B boxes will fit into box A?

   ______

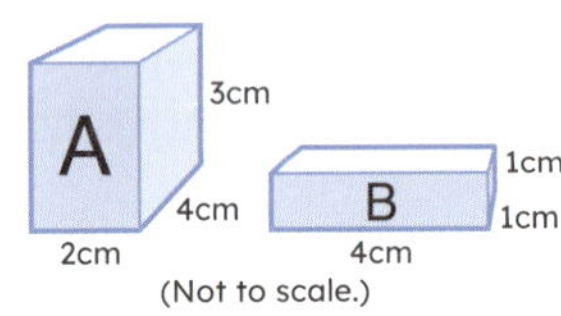

(Not to scale.)

11. 1, 4, 9, ______, 25, 36, ______

12. What is the place value of 2 in 3.812?

   ______

13. 30^2 = ______ × ______ = ______

14. Write one minute past one in the afternoon as 24-hour time.

   ______

15. For a 4m-by-4m bathroom floor, 10 tiles cover $1m^2$. How many tiles are needed to cover the entire floor?

   ______

Tuesday

1. Add brackets to this number sentence.

   3 × 8 – 5 × 2 = 14

2. 30 birds are in the garden. The ratio of galah to magpies is 1:1. How many galah are there?

   ______

3. Use an arrow to show the convex surface.

4. What is the value of 3 in 6.913?

   ______

5. Round 17.73 to the nearest tenth. ______

6. 30 ÷ 5 = 60 ÷ ______

7. 9000 – 1500 = ______

8. If a bus arrived at its final stop at 2:20 pm after four stops with five-minute intervals between each, what time did it start its run?

   ______

9. 40^2 = ______ × ______ = ______

10. What was the GST and the total cost? A farrier's invoice showed:

Labour	$120
Horse shoes	$20
GST @ 10%	______

   Total = ______

11. Name this object.

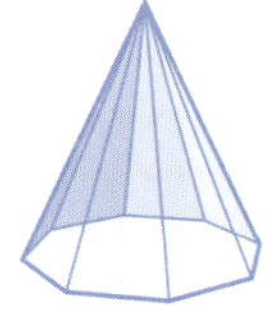

   ______

12. Circle the numbers divisible by 10.

   80 101 201 90 140

13. 1, 5, 25, ______, 625

14. 15% of $10 = ______

15. 8.009 < 8.02 true ☐ false ☐

Wednesday

1. Add brackets to this number sentence.

   6 × 3 + 8 ÷ 2 = 22

2. What is the distance between Orange and Albury? (Note: 10mm = 20km)

   ________km

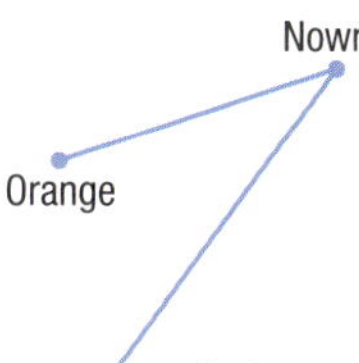

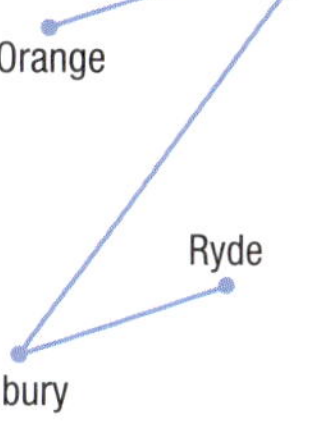

3. What is the distance between Ryde and Nowra?

   ________km

4. $1 + \frac{3}{1000} + \frac{2}{10} + \frac{8}{100} =$ ________ . ________

5. (a) 0.8 + 0.9 = ________

   (b) 0.08 + 0.09 = ________

   (c) 0.008 + 0.009 = ________

6. The LCD for $\frac{1}{8}$ and $\frac{2}{3}$ is ________.

7. Draw a $\frac{3}{4}$ turn clockwise.

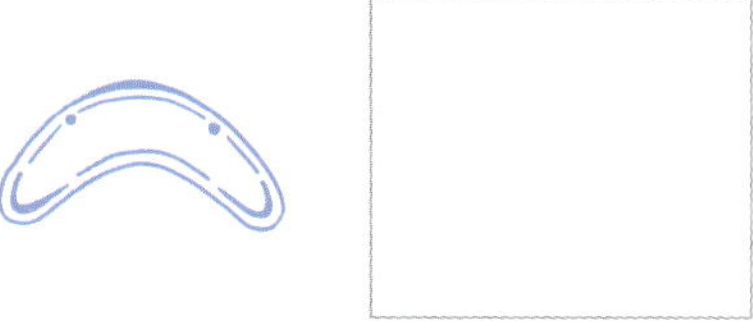

8. If 40 000 – c = 10 000, then c = ________.

9. 10^4 = ________

10. A room is 5m wide by 7m long by 2m high. What is its volume?

    ________m^3

11. $\frac{3}{4}$ = ________%

12. 4 × 6 = 8 × ________

13. $3\frac{4}{5} + 1\frac{3}{5}$ = ________

14. If you rode your bike from A to B, how far did you ride?

    ____

    0 A B 600m

15. If you can ride your bike $2\frac{1}{2}$km in five minutes, how far can you ride in one hour?

    ____

Thursday

1. Add brackets to this number sentence.

   9 × 3 – 18 ÷ 2 = 18

2. 80 × 9 = 720, 79 × 9 = ________

3. How many B boxes will fit into box A?

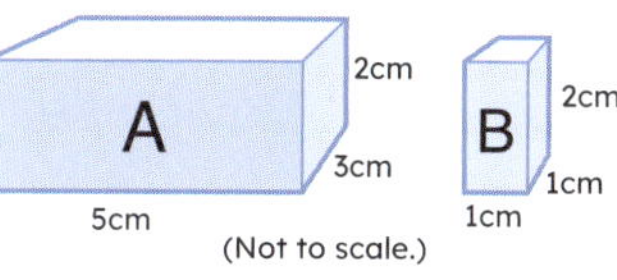

   ____

4. Circle the angle which is closest to 45°.

5. 1 – 0.95 = ________, 0.1 – 0.095 = ________

6. 24 ÷ 4 = ________ ÷ 2

7. 1.85, 1.90, 1.95, ________, 2.05, ________

8. 2^4 = ________

9. You have $6.80, of which $1.80 is 20c coins, and the rest is 50c coins.

   How many coins in total? ________

10. What is the perimeter of the rectangle?

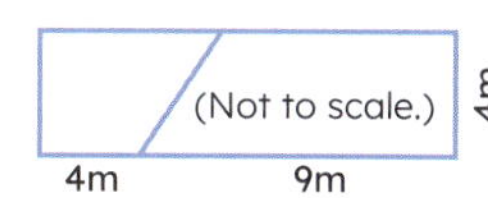

    ____

11. A farm has a square paddock with five fence posts along each side. How many fence posts are there altogether?

    ____

12. True or false?

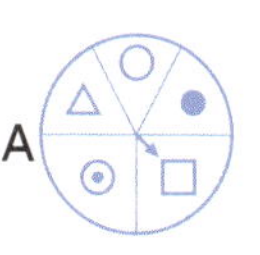

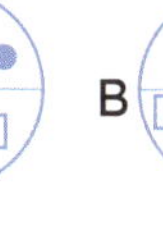

    (a) The chance of spinning a △ on B is greater than △ on A. ________

    (b) The chance of a □ on A is equal to a □ on B. ________

    (c) The chance of a ○ on A is 25%. ________

13. If a rectangular room is 7m by 10m, what is the floor area?

    ________m^2

14. What number is halfway between?

    46 □ 56

15. If a judo class has 27 adults to 9 children what is the ratio?

    ____

Week 7

Week 7

Problem-solving

David bought six apples for 60c each, three bananas for 85c each, and four oranges for 55c each.

Write and solve an equation that uses at least one set of brackets to represent how much David spent.

Read the question again. **Think** about the information. Underline the important words.

Tick the strategy you will use to work out the answer:

- estimate and check ☐
- look for patterns ☐
- draw a diagram or picture ☐
- construct a table or graph ☐
- use materials ☐
- use a formula ☐
- something else. ☐

Solve it:

Reflect on the question and answer.

Check it. Circle another strategy on the list to work it out.

Show it:

Friday Review

1. Add brackets to make this number sentence true.

   7 × 5 − 10 ÷ 2 = 30

2. Halve $\frac{1}{2}$. ________

3. 3 − 0.94 = ________

4. A 3m × 5m floor needs retiling. If eight tiles cover $1m^2$, how many tiles are needed to cover the room?

   ____

5. 15% of $60.00

   = ________

6. What number is halfway between?

   29 ☐ 39

7. What is the volume of a 7m-by-3m-by-2m room?

   ________ m^3

8. What is the chance of spinning a ● on each of these spinners?

   A = ________ in ________

   B = ________ in ________

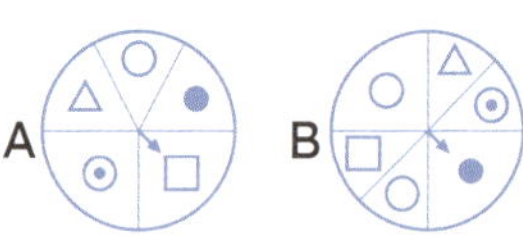

9. Round 23.58 to the nearest tenth.

   ____

10. How many B boxes will fit into box A?

    ____

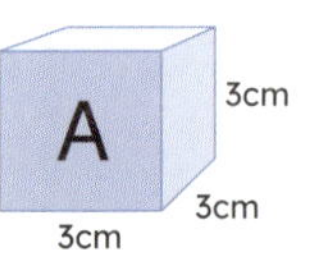

(Not to scale.)

11. If 50 000 − c = 35 000, then

    c = ________.

12. Draw as a $\frac{1}{2}$ turn clockwise.

13. Is 4303 evenly divisible by 10?

    yes ☐ no ☐

14. 2.85, 2.9, 2.95,

    ____

15. Add brackets to this number sentence.

    7 × 8 ÷ 4 × 2 = 7

16. A bus arrives at its final stop at 11:20 pm. It had already stopped four times with eight-minute intervals between each. What time did it leave its depot?

    ____

17. You have $9.60, of which $1.60 is in 20c coins and the rest is in 50c coins. How many coins in total?

    ____

18. What is the radius of a circle with a 15cm diameter?

    ____

St P

978-1-923005-54-9

Monday

1. What is the median score? ______

What is the range? ______

Maths test scores

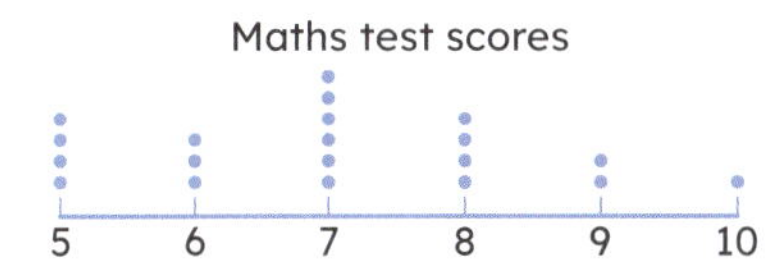

2. 1 – 0.92 = ______, 0.1 – 0.092 = ______

3. 9 × 7 = 100 – ______

4. 9000 – 3500 = ______

5. $3\frac{1}{3} - \frac{2}{3} =$ ______

6. If $c \div 5 = 100$, then $c =$ ______

7. 400 + 700 + 3200 = ______

8. Which two prime numbers have the sum of eight?

______ and ______

9. Which statement is true?

(a) A < B > C ☐ (b) A < B < C ☐

(c) A > B < C ☐ (d) A > B > C ☐

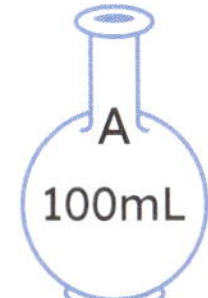

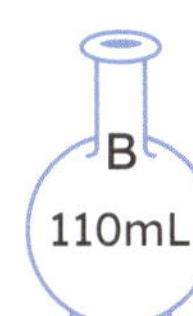

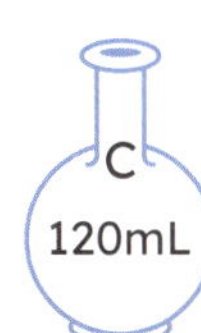

10. Which decimal is closest to 3.18?

(a) 3.176 ☐ (b) 3.19 ☐

(c) 3.2 ☐ (d) 3.185 ☐

11. $11^2 =$ ______

12. 9, 27, 81, ______

13. 100 – 4 × 10 = (a) 96 × 10 ☐

(b) 100 – 40 ☐

14. Which number is divisible by nine?

(a) 2610 ☐ (b) 3571 ☐ (c) 2919 ☐

15. What fraction is shaded? ______

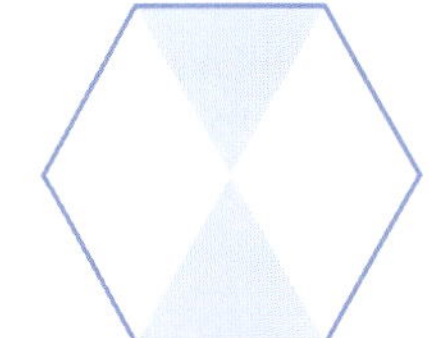

Tuesday

1. Look at Monday's graph.

What is the mean score? ______

What is the mode score? ______

2. What is the angle size of $\angle x + \angle y$?

$\angle x =$ ______

$\angle y =$ ______

$\angle x + \angle y =$ ______

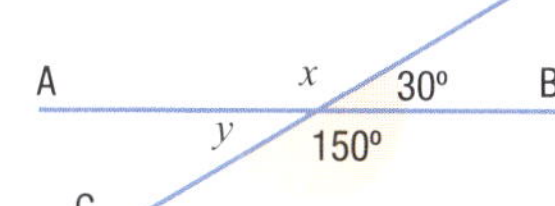

3. 6400 + 1700 = ______

4. What is the perimeter of a block of land 25m long by 20m wide?

______

5. 4 + 5 + 8 = ______

6. $5\frac{1}{5} - \frac{4}{5} =$ ______

7. Draw the grub as a 180° rotation.

8. (9 × 4) ÷ (4 × 3) = ______

9. Name this shape.

______

10. (a) Halve 11. ______

(b) Halve 1.1. ______

11. How many B boxes would fit into box A?

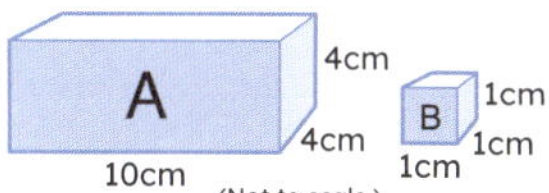

______

12. 22 000 + 19 000 = ______

13. Which two prime numbers have the sum of 12?

______ and ______

14. 15% of 80 is ______.

15. Continue the pattern.

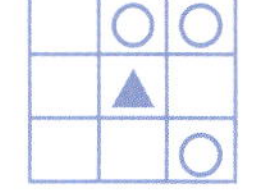
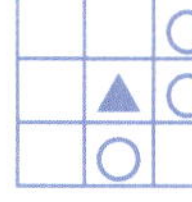
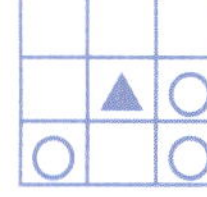

Wednesday

1. What is the median score? ______

What is the range? ______

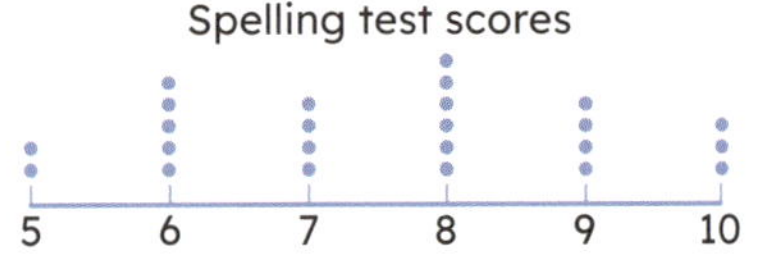

2. If $y \div 4 = 40$, then $y =$ ______.

3. 50 000 − 12 000 = ______

4. 50^2 = ______ × ______ = ______

5. 0.008 + 0.4 + 0.05 = ______

6. Which fraction matches A?

______

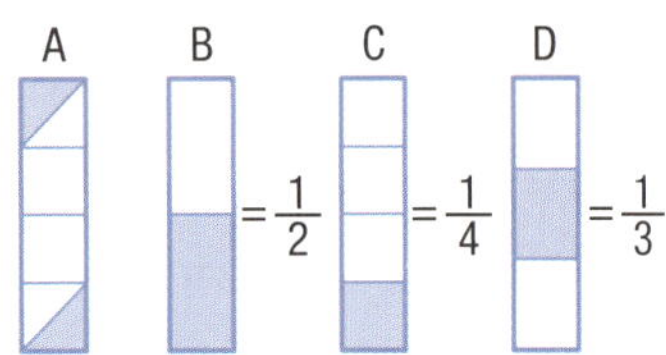

7. Measure $\overline{AB}$. ______mm

8. A square paddock has six fence posts along each side. How many fence posts are there altogether?

______

9. $\frac{2}{3} = \frac{\square}{6} = \frac{\square}{9} = \frac{\square}{12}$

10. $3\frac{7}{10} + 2\frac{7}{10} =$ ______

11. 29 ☐ 100 = 0.29

12. 999 995 + 10 = ______

13. What is the chance of rolling a 2 on a die?

______

14. 5 × 2 + 12 ÷ 6 = ______

15. Shade 1500mL.

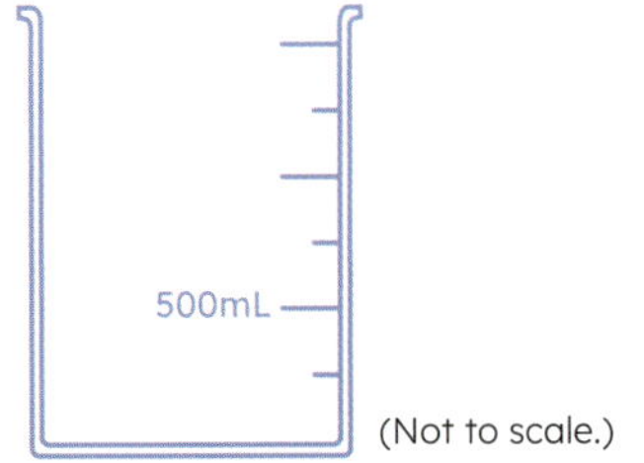

(Not to scale.)

Thursday

1. Look at Wednesday's graph.

What is the mean score? ______

What is the mode score? ______

2. If the time is 11:20 pm, what will it be 26 hours later?

______

3. $\frac{4}{5} = \frac{\square}{15}$

4. What is the place value of 7 in 7 111 233?

______

5. Which two prime numbers have the sum of 16?

______ and ______

6. 25% of 60 is ______.

7. What fraction of the rectangle is shaded?

______

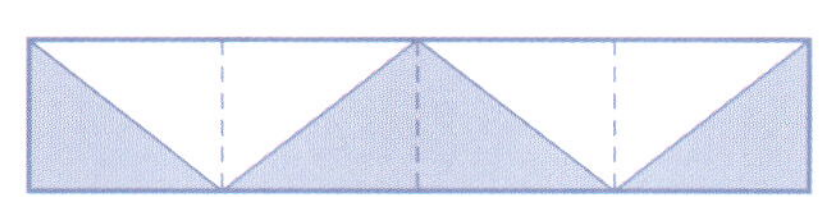

8. 42 − 18 = ______

9. 999 997 + 10 = ______

10. How many B boxes would fit into box A?

______

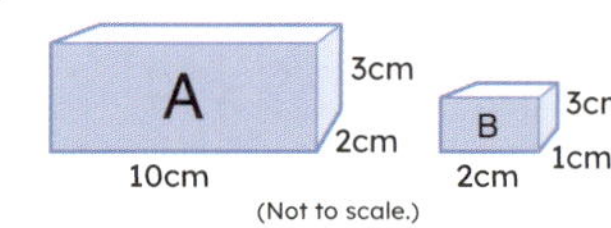

(Not to scale.)

11. $2 + \frac{4}{100} + \frac{8}{10} =$ ______ . ______

12. A book is open and the two page numbers sum to 49. What are the two page numbers?

______ and ______

13. How many 50c coins make up $50.00?

______

14. 3.3, 3.6, 3.9, ______

15. A hat contains 10 yellow balls and 20 green balls. What is the chance of selecting a yellow ball?

______

Problem-solving

The Clark family were planning a trip to Madrid. They could not decide when to visit, so they studied this graph. It shows Madrid's average temperature per month for 12 months.

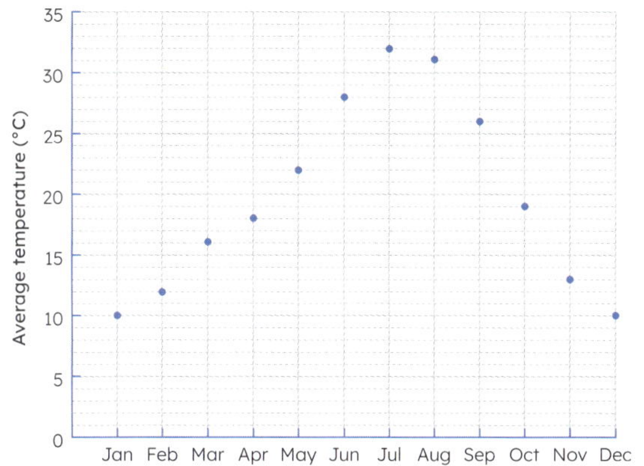

List the mean, mode, median, and range of the monthly average temperatures across the 12 months. Explain why the mode doesn't give an accurate representation of the average temperature.

Read the question again. **Think** about the information. Underline the important words.

Tick the strategy you will use to work out the answer:

- estimate and check ☐
- look for patterns ☐
- draw a diagram or picture ☐
- construct a table or graph ☐
- use materials ☐
- use a formula ☐
- something else. ☐

Solve it:

Reflect on the question and answer.

Check it. Circle another strategy on the list to work it out.

Show it:

Friday Review

1 A group of Scouts went on a camping trip. Their ages were 14, 12, 13, 11, 13, 11, 14, 13, 14, 14, 11, 14, and 12.

(a) What is the age range?

______

(b) What is the median age?

______

(c) What is the mean age?

______

(d) What is the mode age?

______

2 Which two prime numbers, when added, equal 24?

______ and ______

3 10 – 0.92 = ______

4 What fraction of the whole is shaded?

______

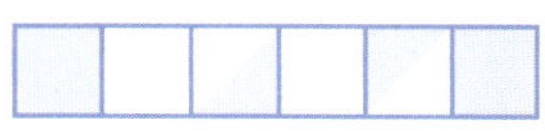

5 If $c \div 3 = 300$, then

c = ______.

6 What is the area of a 50cm-by-20cm sheet of paper?

______

7 2.97, 2.98, 2.99,

______

8 15% of 40 = ______

9 (70 × 10) ÷ (5 × 2)

= ______

10 Which number is divisible by 9?

(a) 3506 ☐

(b) 2511 ☐

(c) 4911 ☐

11 Measure this line in mm.

______mm

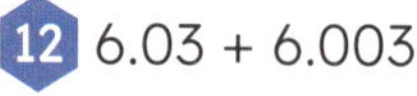

12 6.03 + 6.003

= ______

13 $5\frac{2}{5} - \frac{4}{5}$ = ______

14 Draw a reflection.

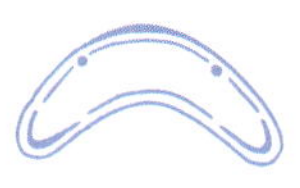

15 What is the chance of randomly picking a king from a deck of 52 playing cards?

______

16 Which statement is true?

(a) A > B < C ☐

(b) A > B > C ☐

(c) A < B > C ☐

A 120mL

B 90mL

C 140mL

17 $5\frac{7}{10} + 3\frac{9}{10}$ = ______

18 How many B boxes will fit into box A?

______

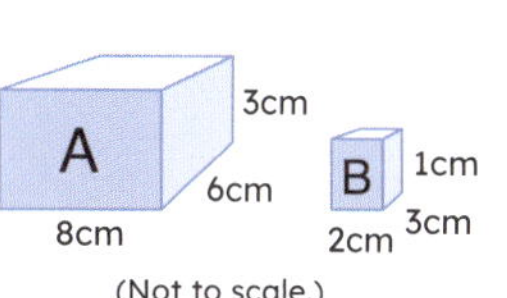

(Not to scale.)

Week 8

Week 9

Monday

1. A triangle has three angles that sum to ______°.

2. 64, 128, 256, ______, 1024

3. 29.35 ÷ 10 = ______

4. If $y \times 300 = 30\,000$, then y = ______.

5. 10 × 700 = ______

6. What number is halfway between?

______

7. What fraction is shaded?

(a) $\frac{1}{4}$ ☐ (b) $\frac{2}{4}$ ☐

(c) $\frac{1}{3}$ ☐ (d) $\frac{1}{2}$ ☐

8. $\frac{1}{8}$ = ______% = 0.______

9. 7 + [5 × (6 ÷ 3)] = ______

(a) 12 × 2 ☐ (b) (7 + 5) × 2 ☐

(c) 17 ☐ (d) 72 ÷ 3 ☐

10. Draw as a 90° clockwise rotation.

11. Double 375. ______

12. Write the angles for the other three corners of this square.

______ ______ ______

90°

13. 8200 + 900 + 600 = ______

14. Write 11:10 pm in 24-hour time. ______

15. A = ______mm

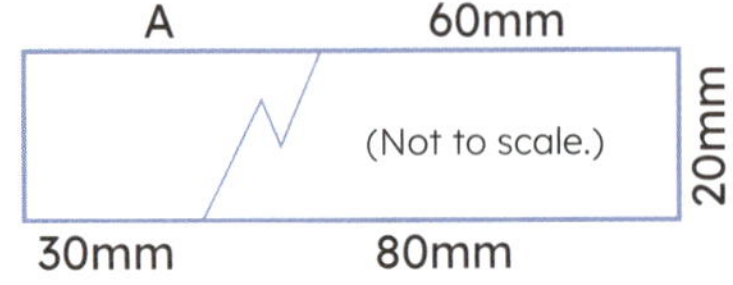

Tuesday

1. Angles a and b both equal ______°.

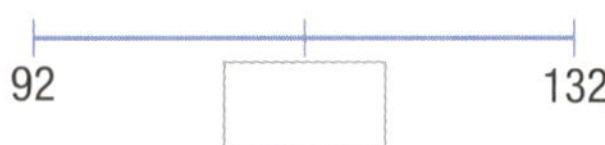

2. What is the number that is halfway between?

92 ☐ 132

3. Halve $\frac{1}{5}$. ______

4. 7^2 = ______

5. $8\frac{8}{10} + 1\frac{2}{10}$ = ______

6. Multiply a number by five, double it, add four, and the answer is 74. What is the starting number?

______

7. Which statement is true?

(a) B > C > A < D ☐

(b) A > D > C > B ☐

(c) A < C < D > A ☐

(d) A > D < C > A ☐

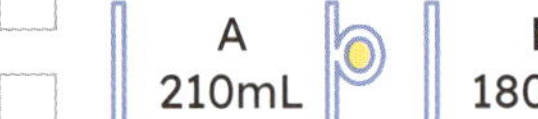

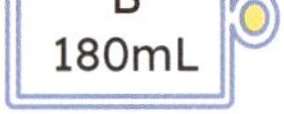

8. $\frac{2}{3} < \frac{3}{8}$ true ☐ false ☐

9. Which is the smallest mass?

(a) 1010g ☐ (b) 1.1kg ☐ (c) 2g ☐

10. 9×10^4 = ______

11. What is the value of 3 in 9.253? ______

12. Volume = ______m^3

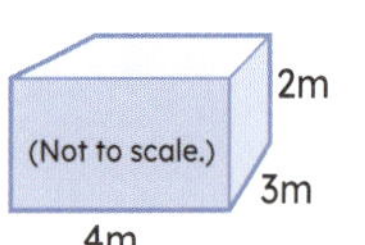

13. 7.1 – 0.2 = ______

14. Record on the chance line.

A jar holds 10 marbles: four red, one blue, and five green. What is the chance of picking a green marble?

15. If you can ride your bike six kilometres in 10 minutes, how far could you go in two hours?

______

Wednesday

1. ∠B = ________°.

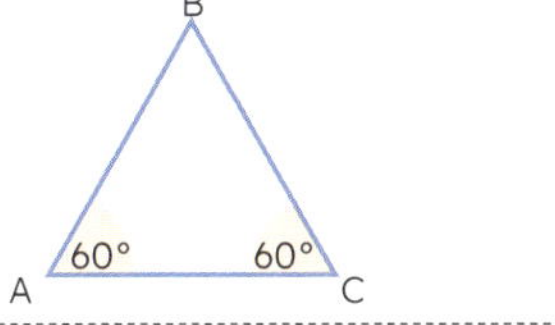

2. Simplify $\frac{20}{50}$. ________

3. 678 + 49 = ________

4. –24 > 5 true ☐ false ☐

5. What is the probability, as a fraction, of a coin landing on tails?

____

6. Draw the diagonals.

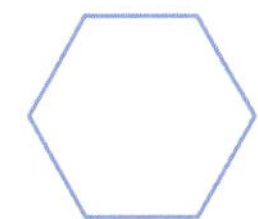

7. If $y \times 700 = 70\,000$, then y = ________.

8. 10 – 0.99 = ________, 1 – 0.099 = ________

9. How many B boxes would fit in box A?

____

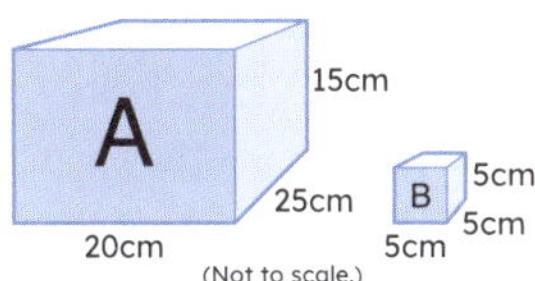

10. 17 + 18 + 15 = ________

11. Multiply a number by four, halve it, subtract nine, and the answer is one. What is the starting number?

____

12. What is the interval between the bus departures?

____

Bus timetable		
	Departs Alex Avenue	12:52
	Arrives 1st Ave	1:07
	Arrives 2nd Ave	1:22
	Arrives 3rd Ave	1:37

13. 999 993 – 8 = ________

14. 15% of $20 is ________.

15. 3 + 4 × 2 – 3 = ________

Thursday

1. A quadrilateral has four angles that sum to ________°.

2. If a bus arrives at its fifth stop at 6:50 am, and its stops are eight minutes apart, what time did it start its route?

____

3. Which number is divisible by three?

(a) 1521 ☐ (b) 2921 ☐ (c) 1483 ☐

4. $\frac{3}{4}$ = 0.________

5. If Lauren scored 70, 45, and 35 in cricket, what is their average (mean) run score?

____

6. 2, 2.9, 3.8, 4.7, ________

7. Name the object.

____

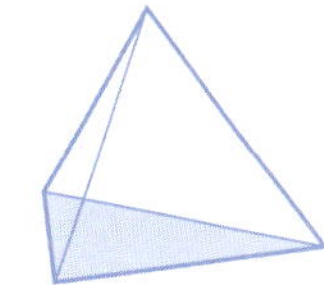

8. 403 – 30 = ________

9. $3\frac{4}{7} + 5\frac{6}{7}$ = ________

10. What is the ratio of students to teachers if there are 12 students and 4 teachers?

____

11. $10.00 – $2.35 = ________

12. 6 + 5 × 2 – 2 = ________

13. Continue the pattern.

14. 6.4 – 0.9 = ________

15. 60^2 = ________ × ________ = ________

Week 9

Problem-solving

Tim and Gina want to tile their bathroom floor using hexagonal tiles. First, they must work out the size of the missing angles.

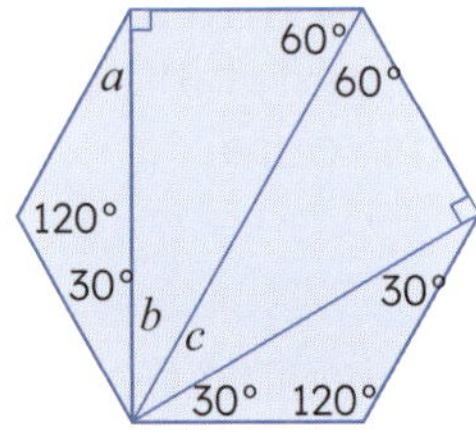

What are the sizes of the angles a, b, and c?

Read the question again. **Think** about the information. Underline the important words.

Tick the strategy you will use to work out the answer:

- estimate and check ☐
- look for patterns ☐
- draw a diagram or picture ☐
- construct a table or graph ☐
- use materials ☐
- use a formula ☐
- something else. ☐

Solve it:

Reflect on the question and answer.

Check it. Circle another strategy on the list to work it out.

Show it:

Friday Review

1 Which regular polygon has 120° internal angles?

(a) square ☐

(b) pentagon ☐

(c) hexagon ☐

(d) octagon ☐

2 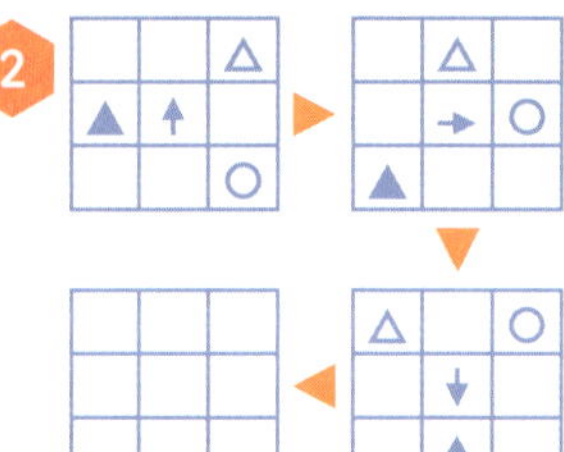

3 What is the place value of 5 in 2.965?

____

4 Double 570. ________

5 If you ride your bike 6km in 12 minutes, how far can you ride in two hours?

____

6 What is the perimeter?

____

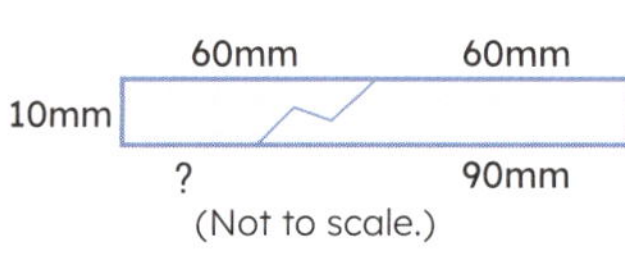

(Not to scale.)

7 The angles of a parallelogram add up to ________ °.

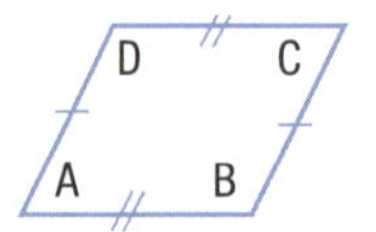

8 Simplify $\frac{15}{24}$. ________

9 If the time is 10:35 pm, what will it be 27 hours later?

____

10 If y × 700 = 70 000, then y = ________.

11 72 000 – 29 000 = ________

12 $7\frac{5}{7} + 4\frac{6}{7} =$ ________

13 What is the chance of you riding a horse to school tomorrow?

unlikely ☐

likely ☐

impossible ☐

certain ☐

14 Draw as a 90° clockwise rotation.

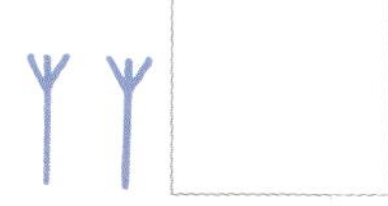

15 6 × 9 = (7 × 9) – ________

16 $1\frac{1}{4}$ = ________%

17 What is the volume of a 15m × 4m × 2m sea container?

____

18 80^2 = ________ × ________

= ________

N A M Sp St P

Monday

1. If you flip two coins simultaneously, what is the probability of landing one heads and one tails?

   ________ in ________

2. 7500 + 5500 + 6500 = ________

3. Three vertically stacked cubes are painted. Once separated, how many cube faces are not painted?

   ____

4. A fast train travelled 600km in three hours. What was the average speed?

   ________km/h

5. 20 × 3 = 10 × ________

6. 9 × 8 = (10 × 8) – ________

7. Label the remaining angles.

   a = ________°

   b = ________°

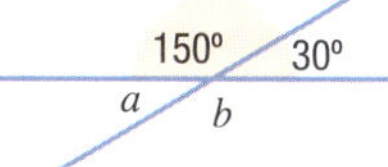

8. The perimeter of a regular pentagon with 9cm sides is

   ________.

9. 50 + 90 = 2 × ________

10. odd + odd + odd = ________

    odd + even + odd = ________

    even + even + odd = ________

11. Which decimal is closest to 7.79?

    (a) 7.8 ☐ (b) 7.78 ☐

    (c) 7.789 ☐ (d) 7.792 ☐

12. 1, 1.8, 2.6, 3.4, ________

13. If Alexander swam 50m in 54.2 seconds and Antonio's time was 53.8 seconds, how many seconds faster was Antonio?

    ____

14. 70 ÷ 7 × 2 + 4 = ________

15. What fraction of these shapes are triangular?

    ____

Tuesday

1. If you flip two coins simultaneously, what is the probability of landing two heads?

   ________ in ________

2. 400 – 150 = ________

3. How many 20c coins make up $11.00?

   ____

4. Write the factors of 12. ________

5. What is the LCD for $\frac{2}{5}$ and $\frac{1}{4}$? ________

6. What is the ratio of roses to daisies if there are 21 roses and seven daisies?

   ____

7. A jar contains five green marbles, four glass marbles, and 11 red marbles. What is the probability of choosing a glass marble?

   ____

8. Draw the reflection.

9. Does 1297 + 1347 + 1433 equal an odd or even number?

   ____

10. Double 2.7. ________

11. Draw as a 270° clockwise rotation.

12. Measure the length of this line. ________mm

13. Round π (3.14) to the nearest whole.

    ____

14. 9 × 8 = (5 × 8) + (4 × 8) = ________

15. Jin swam 50m in 53.9 seconds, and 54.2 seconds. What is the time difference?

    ____

Week 10

Wednesday

Week 10

1. If you flip two coins simultaneously, what is the chance of them both landing on tails?

   ________ in ________

2. What is the place value of 2 in 8.712?

   ____

3. If $400 \div y = 50$, then $y =$ ________.

4. What is the chance of randomly picking a red-coloured card from a deck of 52 playing cards?

   ________%

5. $\frac{1}{3} = 0.333 =$ ________%

6. 4500 + 3500 + 8500 = ________

7. Halve 4.9. ________

8. Does 1698 + 3576 + 1783 equal an even or an odd number?

   ____

9. A fast train travels 300km in two hours. What is its average speed?

   ________km/h

10. (200 × 10) – (5 × 10) = ________

11. Farmer Mae has a square paddock with six fence posts on each side. How many fence posts are there altogether?

    ____

12. Which side is heavier?

    ____

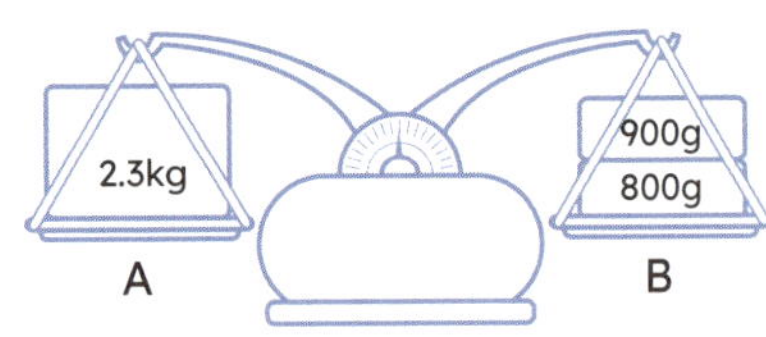

13. 100 000 – 32 000 = ________

14. 60.5 ÷ 5 = ________

15. If you can ride your bike 11km in 30 minutes, how far could you ride in three hours if you rode at the same speed without stopping?

    ____

Thursday

1. If you flip three coins simultaneously, what is the chance of them all landing on heads?

   ________ in ________

2. It is 2:25 am on a Tuesday. What will the time and day be 48 hours later?

   ____

3. 12 × 8 = (10 × 8) + (2 × 8) = ________

4. $90^2 =$ ________ × ________ = ________

5. Round 12.27 to the nearest tenth. ________

6. Simplify $\frac{30}{40}$. ________

7. 3, 3.7, 4.4, ________, 5.8

8. $17 \times 10^4 =$ ________

9. 12 ÷ 3 × 4 = ________

10. Write the factors of 24.

    ____

11. How many degrees are in a circle? ________

12. What is the size of angle a?

    ____

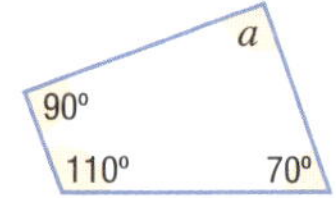

13. 22 × 50 = (22 × 100) ÷ 2 = ________

14. Coen swam 100m in 59.4 seconds, Jedda's time was 58.7 seconds, and Anthony's 58.9 seconds. Order the swimmers from first to third.

    1. ________
    2. ________
    3. ________

15. Double 4.6. ________

Problem-solving

When you roll a standard die, the chance of it landing on a six is 1 in 6. If you roll two standard dice simultaneously, what is the chance of rolling two sixes?

Read the question again. **Think** about the information. Underline the important words.

Tick the strategy you will use to work out the answer:

- estimate and check ☐
- look for patterns ☐
- draw a diagram or picture ☐
- construct a table or graph ☐
- use materials ☐
- use a formula ☐
- something else. ☐

Solve it:

Reflect on the question and answer.

Check it. Circle another strategy on the list to work it out.

Show it:

Friday Review

Week 10

1. If you throw two dice simultaneously, what is the probability of throwing two numbers less than two? ______

2. Which decimal is closest to 2.31?
   (a) 2.32 ☐
   (b) 2.3 ☐
   (c) 2.308 ☐
   (d) 2.311 ☐

3. Circle the side that will tip the balance.

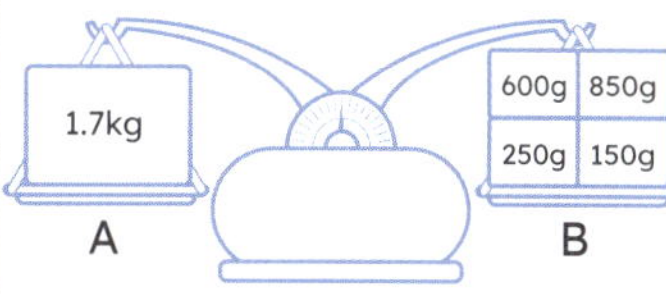

4. If you can walk 1km in 15 minutes, how far could you walk in two hours? ______

5. 25% of 10 is ______.

6. 3.7 − 0.9 = ______

7. Round π to the nearest whole. ______

50m swim times
Lola – 52.6 sec.
Mia – 52.8 sec.
Allira – 53.7 sec.

8. What is the time difference between first and third place? ______

9. What is the range of the data in the table? ______

10. If $y \times 50 = 3000$, then $y =$ ______.

11. 2.6, 3.2, 3.8, ______

12. What is the perimeter of a regular hexagon with 6cm sides? ______

13. If you flip three coins simultaneously, what is the chance of them all landing on tails? ______ in ______

14. What is the LCD for $\frac{2}{5}$ and $\frac{3}{4}$? ______

15. $a° = b° =$ ______

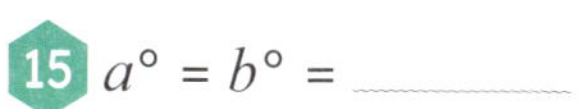

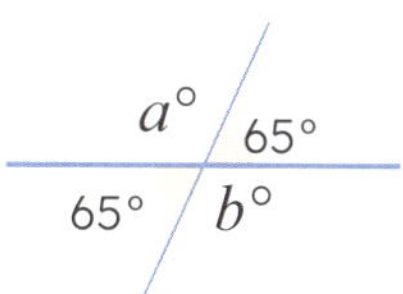

16. 30 ÷ 5 + 5 = ______

17. Draw as a 180° clockwise rotation.

18. 999 997 + 13 = ______

N A M Sp St P

Week 11

Monday

1. Nicholas read several books. Alana read four more books than Nicholas. Which expression can be used to show the number of books Alana read? (n = Nicholas)

   (a) $n - 4$ ☐ (b) $n + 4$ ☐
   (c) $n \times 4$ ☐ (d) $n \div 4$ ☐

2. Draw the front view.

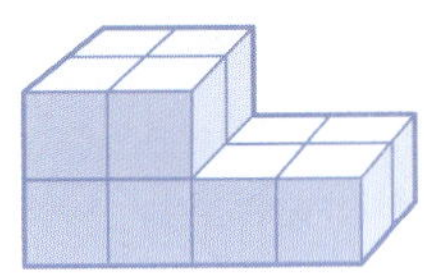

3. Write the number 101 101 in words.

   ____

4. 1.37 = 1 + 0.3 + ______

5. Which decimal is located between $\frac{1}{3}$ and $\frac{1}{2}$?

   (a) 0.5 ☐ (b) 0.4 ☐ (c) 0.6 ☐
   (d) 0.32 ☐ (e) 0.23 ☐

6. This has rotational symmetry to the order of

   ______.

7. 290 + 150 = ______

8. If the Australian Western Standard Time (AWST) during winter is 6 pm, what time is it in New South Wales (NSW)?

   ______

9. Double 1850. ______

10. Measure this line. ______mm

11. Can a regular hexagon tessellate?

    yes ☐ no ☐

12. 6^2 = ______

13. 10% of \$50.00 = ______

14. 6 + 6 + 6 + 6 = 4 × 6 = ______

15. 1.25 =

    (a) $1\frac{2}{3}$ ☐ (b) $1\frac{1}{4}$ ☐ (c) $1\frac{1}{2}$ ☐

Tuesday

1. During a tomato eating contest, Thomas ate six less than Daniel. Which expression can be used to show the number of tomatoes that Daniel ate? (t = Thomas)

   (a) $t + 6$ ☐ (b) $t - 6$ ☐
   (c) $t \times 6$ ☐ (d) $t \div 6$ ☐

2. $1\frac{1}{3}$ =

   (a) 1.3 ☐ (b) 1.33 ☐ (c) 1.13 ☐

3. Andrew scored an average (mean) of 65 runs over six innings of cricket. Total the number of runs scored.

   ______

4. 14 − 7 = ______, 140 − 70 = ______

5. Jan paid \$450 for some shoes, and Gloria paid 10% less for the same style shoes. How much was the cheaper price?

   ______

6. 4, 14, 34, 64, ______

7. 69 + 43 = ______

8. Draw the right side view.

9. 55 ÷ 5 = 110 ÷ ______ = ______

10. If a car travels 6km in 10 minutes, how far can it travel in one hour at the same speed?

    ______

11. Can a square tessellate? yes ☐ no ☐

12. Arrange these digits to form the largest value: 6, 7, 0, 1, 5.

    ______

13. \$50.00 − \$4.50 = ______

14. Order these fractions from smallest to largest:

    $\frac{1}{2}$ $\frac{1}{5}$ $\frac{4}{5}$ $\frac{2}{3}$ ______ ______ ______ ______

15. $\frac{1}{2} + \frac{1}{4} = \frac{\square}{4}$

Wednesday

1. In netball, Olivia scored five more goals than Mia. Which expression can be used to show the number of goals Olivia scored? (m = Mia)

   (a) $m + 5$ ☐ (b) $m - 5$ ☐

   (c) $m \times 5$ ☐ (d) $m \div 5$ ☐

2. Which two prime numbers, when added, equal 40?

   ________ and ________

3. $6 \times 9 =$ ________

4. Draw the top view.

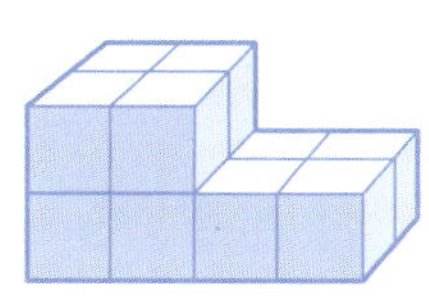

5. A cricket team's batting scores are: 10, 30, 35, 30, 1, 0, 51, 30, 9, 30. What is the mode?

   ____

6. $29.254 \times 100 =$ ________

7. If AWST is midnight, what time is it in Queensland? ________

8. $\frac{2}{3} + \frac{1}{6} = \frac{\square + \square}{6} = \frac{\square}{6}$

9. $\frac{1}{5} = 0.$________

10. $\frac{3}{10} = 0.$________ $=$ ________%

11. $1\frac{3}{4} =$ (a) 1.25 ☐ (b) 1.75 ☐

12. Round 2.37 to the nearest ten. ________

13. Measure this line. ________mm

14. If a car travels 8km in 10 minutes, how far will it travel in one hour at the same speed?

    ____

15. Simplify $\frac{21}{24}$. ________

Thursday

Week 11

1. Kim sang several songs at karaoke. Lucy sang two more songs than Kim. Which expression can be used to show the number of songs Lucy sang? (k = Kim)

   (a) $k + 2$ ☐ (b) $k - 2$ ☐

   (c) $k \times 2$ ☐ (d) $k \div 2$ ☐

2. Over six days, Constable Colin caught 108 motorists speeding. What is the average (mean) number of drivers caught each day?

   ____

3. Draw a double of this square. (Show measurements.)

4. By how many times has the area of the square been increased?

   (a) 12 ☐ (b) 4 ☐ (c) 6 ☐

5. The ratio of mobile phones to students is 1:3. If there are 600 students, how many phones are there?

   ____

6. $-6 > -4$ true ☐ false ☐

7. $1.25 =$ (a) $1\frac{2}{3}$ ☐ (b) $1\frac{1}{4}$ ☐ (c) $1\frac{1}{2}$ ☐

8. What is the median of these numbers: 4, 6, 2? ________

9. If $70 + y = 140$, then $y =$ ________.

10. Angle AB it must add up to

    ________.

    A B

11. If you want a △, which tub is more likely to be chosen to pick from when blindfolded?

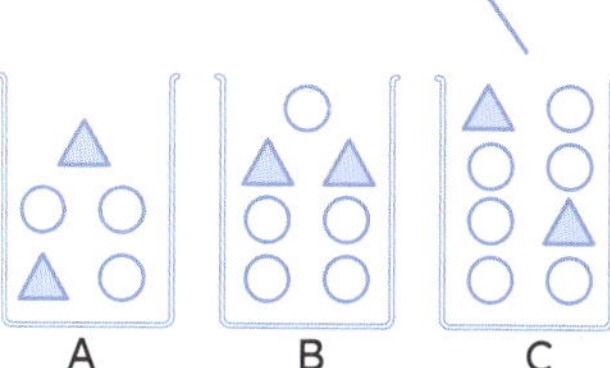

    ____

12. 16, 32, 64, 128, ________

13. A dodecagon has ________ sides.

14. Complete the sequence.

    0.________, 0.________, 0.________, 8, 80, 800

15. If $60 + 40 = 50 + y$, then $y =$ ________.

Problem-solving

Class 7 had rehearsals from Monday to Thursday. Each rehearsal day, they practised for two hours in the morning and then practised again in the afternoon. Every afternoon rehearsal lasted the same amount of time. In total, they practised for 16 hours.

Which of these equations can you use to find the number of hours each afternoon rehearsal took (represented by a)? Circle then solve.

$2a + 4 = 16$ $2(a + 4) = 16$

$4(a + 2) = 16$ $4a + 2 = 16$

Read the question again. **Think** about the information. Underline the important words.

Tick the strategy you will use to work out the answer:

- estimate and check ☐
- look for patterns ☐
- draw a diagram or picture ☐
- construct a table or graph ☐
- use materials ☐
- use a formula ☐
- something else. ☐

Solve it:

Reflect on the question and answer.

Check it. Circle another strategy on the list to work it out.

Show it:

Friday Review

1. Sam had eight less stickers than Alice. Which expression can be used to show the number of stickers that Sam had? (a = Alice)
   (a) $a + 8$ ☐
   (b) $a - 8$ ☐
   (c) $a \times 8$ ☐
   (d) $a \div 8$ ☐

2. Which two prime numbers, when added, equal 30?
   ________ and ________

3. If $40 + y = 130$, then
   y = ________.

4. $\frac{3}{5} + \frac{3}{5}$ = ________

5. 10% of $30 = ________

6. Arrange these digits to form the largest value: 4, 9, 7, 5, 1.
   ____

7. Can a circle tessellate?
   ____

8. Measure line $\overline{AB}$.
   ________ mm
   (line from A to B)

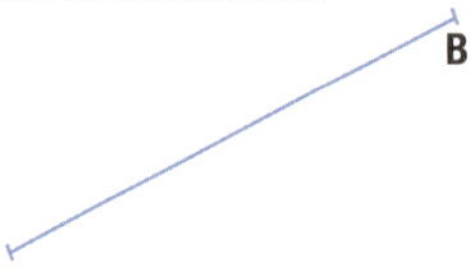

9. If a car travels 7km in 10 minutes, how far can it travel in one hour?
   ____

10. 85 ÷ 5 =
    ________ ÷ 10 = ________

11. Which decimal is between $\frac{1}{4}$ and $\frac{1}{2}$?
    (a) 0.3 ☐ (b) 0.2 ☐
    (c) 0.6 ☐ (d) 0.03 ☐

12. What is the mean (average) of these class results (out of 20): 18, 6, 8, 14, 4?
    ____

13. Draw a front view.

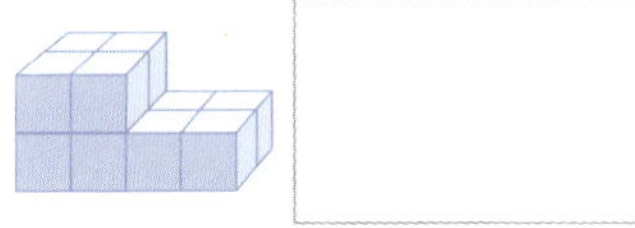

14. Write $3\frac{1}{4}$ as a decimal.
    ____

15. Halve 3958. ________

16. odd + even + odd
    = ________

17. Order the tubs from most likely to have a ▲ chosen to least likely.
    ____

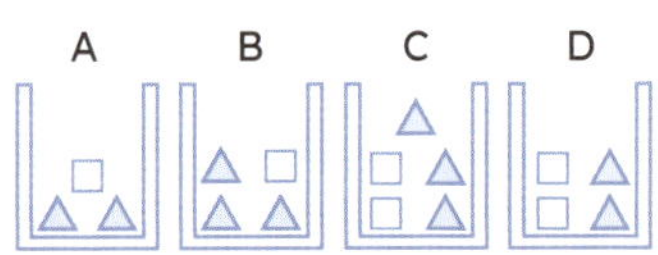

18. Kieran kicked two more goals than Craig. Which expression matches the number of goals Craig kicked? (k = Kieran)
    (a) $k + 2$ ☐
    (b) $k - 2$ ☐
    (c) $k \times 2$ ☐
    (d) $k \div 2$ ☐

P

Monday

1. This is a net of a

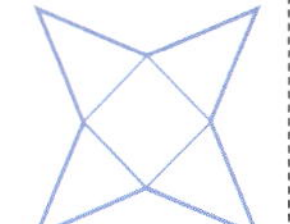

______________________.

2. A farm measuring 500m in length by 400m in width has an area of __________ m^2.

3. 999 993 + 8 = __________

4. Three cubes are painted. Once separated, how many cube faces are not painted?

____

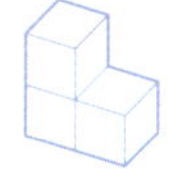

5. 20 × 3 = 10 × __________

6. Which fraction is halfway between $\frac{1}{4}$ and $\frac{1}{2}$?

____

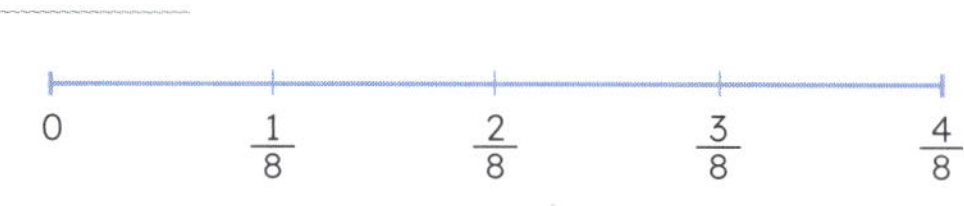

7. In a box of 39 marbles, the ratio of pink to purple is 2:1. How many pink marbles are there?

____

8. For a class test, the marks were 80%, 70%, and 60%. What is the mean score?

____

9. 50 + 90 = 2 × __________ = __________

10. $2\frac{1}{5}$ =

(a) 2.15 ☐ (b) 2.2 ☐

11. Change $\frac{17}{5}$ to a mixed number. __________

12. 250, 500, 750, __________, 1250, __________, 1750, 2000

13. What number is halfway between 105 and 125?

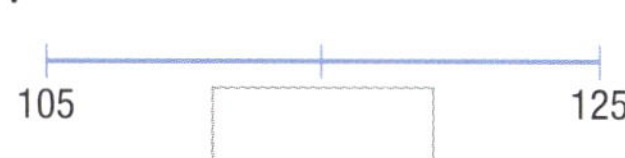

14. $10^3 + 1$ = __________

15. What is the time difference?

____

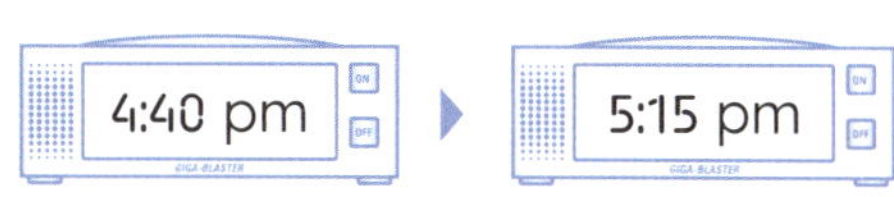

Tuesday

1. This is a net of a

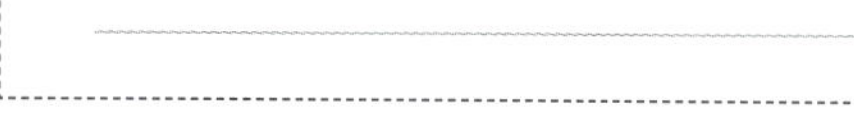

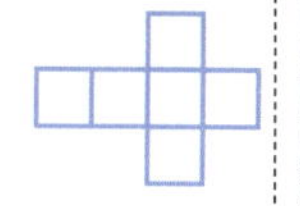

______________________.

2. \$50.00 – \$6.50 = __________

3. 48 + 57 = __________

4. From which tub are you more likely to pick a white marble?

____

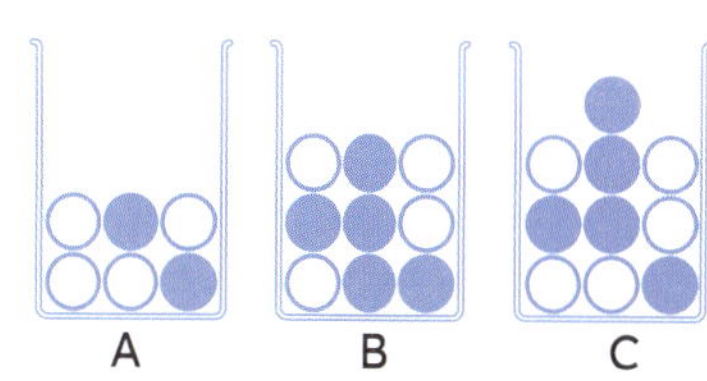

5. $2\frac{3}{5}$ =

(a) 2.3 ☐ (b) 2.6 ☐

(c) 2.35 ☐ (d) 2.15 ☐

6. 16 – 7 = __________, 160 – 70 = __________

7. 0.7 = $\frac{7}{10}$ = __________%

8. Draw the top view of this 3D object.

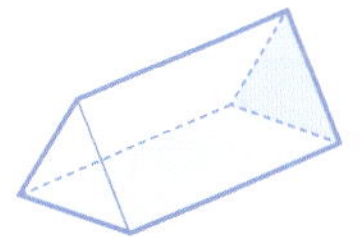

9. Double 1550. __________

10. If a blue car travels 10km in six minutes, how far could it travel in one hour?

____

11. Draw to show a 450° clockwise rotation.

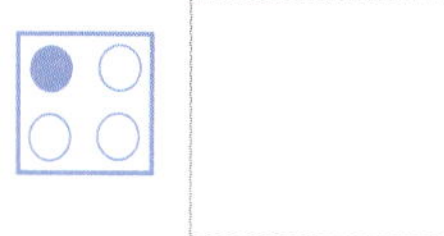

12. 20% of \$80 = __________

13. $\frac{3}{4}$ of 56 = __________

14. If 3 × 10 = 5 × y, then y = __________.

15. b = __________°

c = __________°

a = __________°

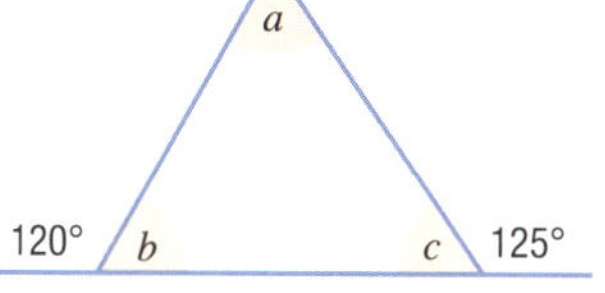

Wednesday

1. What shape does the net make?

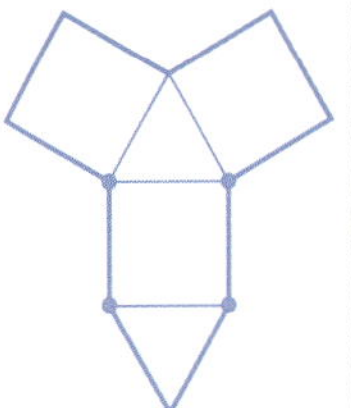

2. $6\frac{3}{5} + 2\frac{2}{5} =$ ________

3. 3, 5, 8, 12, ________, 23

4. Which fraction is halfway between $\frac{1}{3}$ and $\frac{2}{3}$? ________

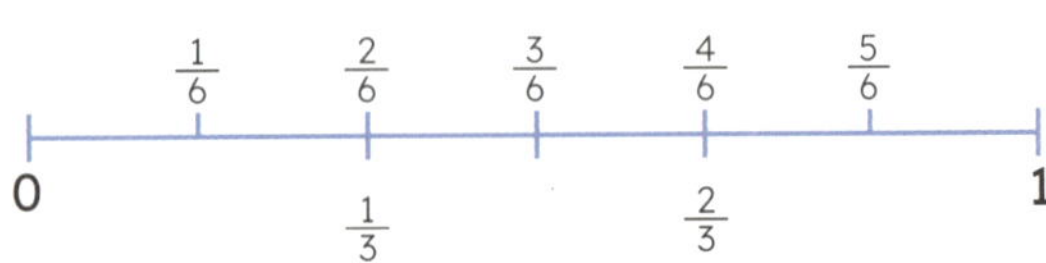

5. Draw a front view.

6. $\frac{\square}{10} = 0.9$

7. Write the median of 6, 8, 10, and 12.

____

8. What size is angle a? ________

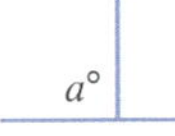

9. Perimeter is to rectangle as circumference is to ________.

10. $24 \times 16 =$ ________

11. $3 - 0.03 =$ ________, $0.3 - 0.003 =$ ________

12. $63 + 19 =$ ________

13. What is the sum of the internal angles for a:

(a) square? ________

(b) pentagon? ________

(c) hexagon? ________

14. If $80 + 40 = 60 \times y$, then $y =$ ________.

15. What is the time difference? ________

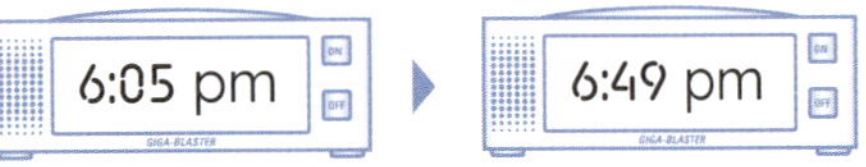

Thursday

1. This is a net of a

____

________.

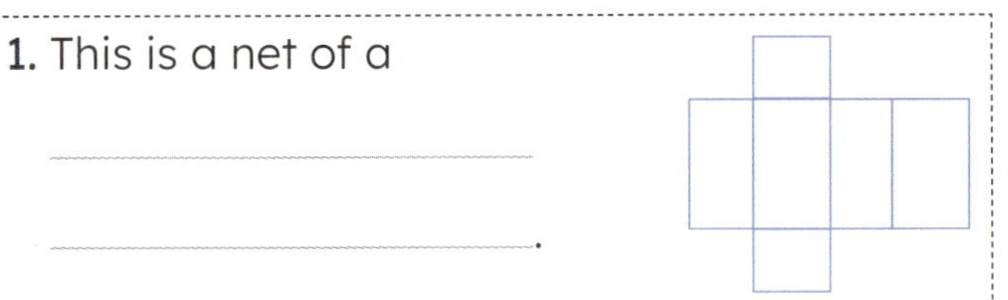

2. 9, 11, 13, 11, 2, 5, 12, 11

The mode is ________. The median is ________.

3. Halve 280. ________

4. Draw a shape double the size of this rectangle and show the measurements.

5. By how many times has the area of the shape increased?

(a) 2 ☐ (b) 4 ☐ (c) 6 ☐

6. $27 \times 12 = (20 + 7) \times (10 + 2) =$ ________

7. $999\,997 + 9 =$ ________

8. What is the angle size of a?

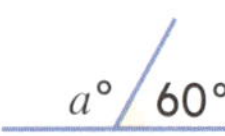

____

9. $\frac{3}{4} - \frac{1}{2} = \frac{\square}{4}$

10. What is the time difference? ________

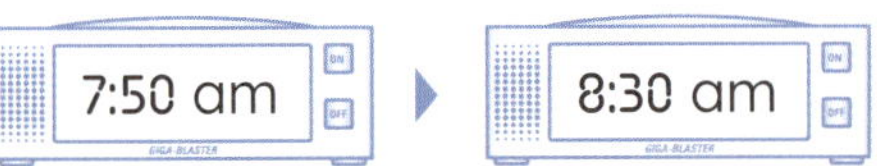

11. $\frac{2}{3} + \frac{5}{6} = \frac{4}{6} +$ ________

12. If a car is travelling at 50km/h, how far could it travel in 30 minutes?

____

13. What is the angle size of a?

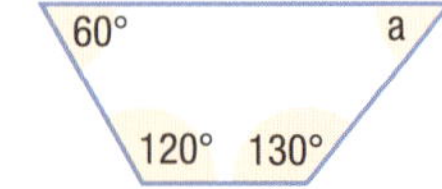

____

14. Round 6.08 to the nearest tenth. ________

15. $5\frac{4}{5} =$ (a) 5.8 ☐ (b) 5.45 ☐ (c) 5.54 ☐ (d) 10.4 ☐

Problem-solving

Consider the faces and edges on triangular, square, pentagonal, and hexagonal prisms.

What patterns do you notice between each prisms' 2D faces and the number of edges on the 3D object?

Read the question again. **Think** about the information. Underline the important words.

Tick the strategy you will use to work out the answer:

- estimate and check ☐
- look for patterns ☐
- draw a diagram or picture ☐
- construct a table or graph ☐
- use materials ☐
- use a formula ☐
- something else. ☐

Solve it:

Reflect on the question and answer.

Check it. Circle another strategy on the list to work it out.

Show it:

Friday Review

1 This is a net of a

____

______________.

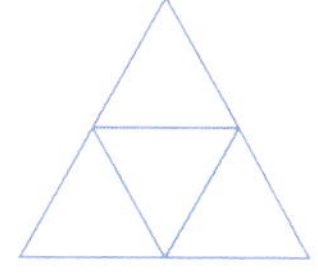

2 20% of $70.00 is

__________.

3 What is the mode of 10, 12, 11, 10, 13, and 15?

____

4 Write $4\frac{2}{5}$ as a

decimal. __________

5 $0.3 = \frac{\square}{10}$

6 $50.00 – $8.50

= __________

7 What is the time difference?

____

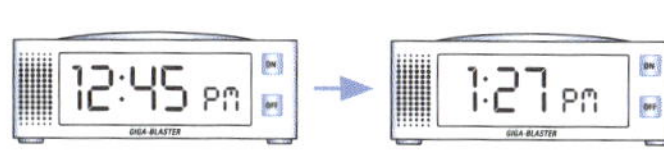

8 5 – __________ = 4.96

9 $10^3 + 9 =$ __________

10 If you turn the rectangle 180° clockwise, which is its new position?

____

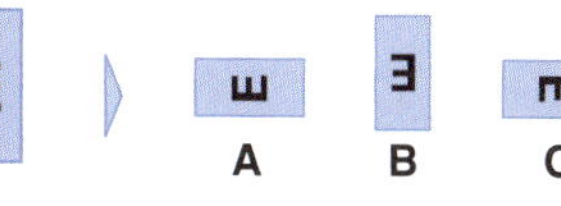

11 This is a net of a

______________.

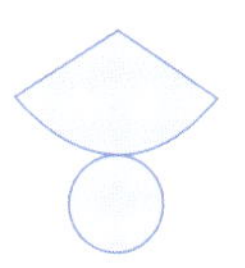

12 $\frac{3}{4} - \frac{1}{2} =$ __________

13 999 997 + __________

= 1 000 006

14 Three same-sized cubes are vertically stacked. Draw the top view.

15 In a box of 42 marbles, the ratio of red to green is 2:1. How many green marbles are there?

____

16 Draw to show a 540° clockwise rotation.

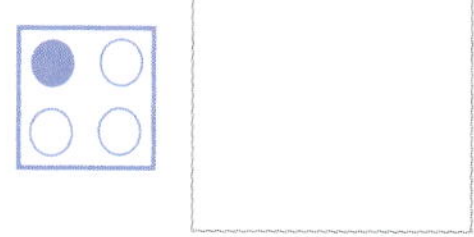

17 How many degrees is $a°$?

____

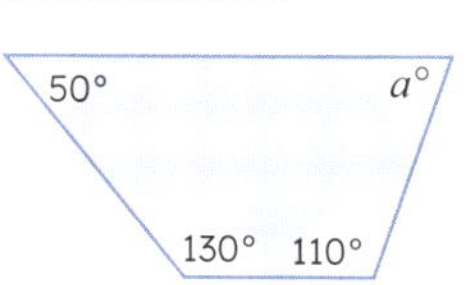

18 If a car travels 11km in six minutes, how far will it travel in one hour?

____

Week 13

Monday

1. 100 = ________ × ________ = 10^2

2. If 100 × a = 1000, then a = ________.

3. 5500 + 7500 = ________

4. If ⬆ is north, what direction is A to B?

(a) NE ☐ (b) NW ☐

(c) SE ☐ (d) SW ☐

5. 16 ÷ 5 = ________

6. 1% = 0.01 = $\frac{\square}{100}$

7. 0.07, 0.08, 0.09, ________

8. Which face on the cube net is opposite to E?

____

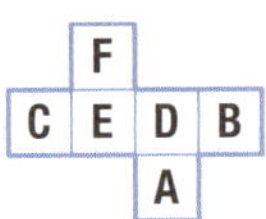

9. Area of a circle = π × ________

10. How many 50c coins make up $20.00?

____

11. 46 × 35 = (40 + ________) × (30 + ________)

= (40 × ________) + (6 × ________)

= (40 × ________) + (6 × ________)

12. A is at which coordinate?

(a) (2,3) ☐ (b) (3,2) ☐

(c) (5,1) ☐ (d) (2,2) ☐

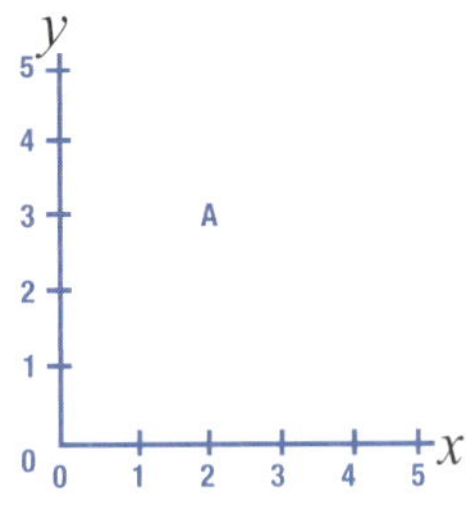

13. 2000 – 350 = ________

14. Round 6.073 to the nearest hundredth.

____

15. 8.08 > 8.1

true ☐ false ☐

Tuesday

1. 1000 = ________ × ________ × ________ = 10^3

2. If it is 38°C, which season is it likely to be?

____

3. If $\frac{2}{3} + y = 1$, then y = ________.

4. 64, 16, 4, ________, ________

5. Write one hundred and eleven thousand and eleven as a numeral.

____

6. Draw the top view.

7. 2% = 0.02 = $\frac{\square}{100}$

8. $3\frac{3}{5}$ = ____.________

9. 4 – 0.007 = ________

10. Measure $\overline{XY}$. ________mm

11. If ⬆ is north, what direction is B to A?

(a) NE ☐ (b) NW ☐

(c) SE ☐ (d) SW ☐

12. Multiply a number by 6, halve it and the answer is 15. What is the starting number?

____

13. Halve $\frac{1}{2}$. ________

14. 10^5 = ________

15. Circle the prime numbers.

1, 2, 3, 4, 5, 6, 7, 8, 9

Wednesday

1. 10 000 = ______ × ______ × ______ × ______ = 10^4

2. Alicia (a) has 7 more songs on their phone than Jin (j) does. Which expression shows the amount of songs Jin has?

   (a) $a + 7$ ☐ (b) $a - 7$ ☐

   (c) $j + 7$ ☐ (d) $j - 7$ ☐

3. 40% of $400.00 = ______

4. y = ______°

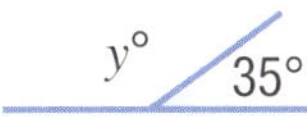

5. 18 ÷ 5 = ______

6. $\frac{1}{3}$ = 0.______

7. 5% = $\frac{\square}{100}$ = 0.05

8. 2 + 0.003 + 0.7 + 0.08 = ______

9. What fraction is halfway between $\frac{1}{5}$ and $\frac{2}{5}$?

______

10. If $y + \frac{6}{10} = 1$, then y = ______.

11. A quadrilateral has 4 internal angles that add up to

    ______°.

12. 999 996 + 9 = ______

13. Label each shape.

    a = pentagon

    b = hexagon

    c = octagon

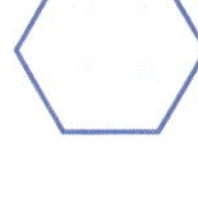

______ ______ ______

14. $\frac{17}{6}$ = ______

15. List the prime numbers < 10.

______

Thursday

1. 100 000 = ______ × ______ × ______ × ______ × ______ = 10^5

2. What is the diameter of this circle?

   ______ cm

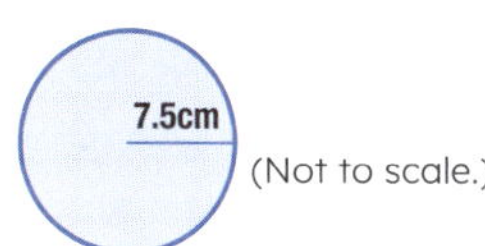

3. A chef measured 9 one-third cups of flour.

   This is equivalent to ______ whole cups.

4. Round 4.075 to the nearest hundredth.

   ______

5. If $1\frac{1}{2} + y = 3\frac{1}{2}$, then y = ______.

6. Simplify $\frac{12}{15}$. ______

7. $\frac{9}{100}$ = 0.09 = ______%

8. In a pie chart, graph Room 1's data about favourite animals: lion – 10%; zebra – 5%; lemur – 30%; koala – 40%; bat – 15%.

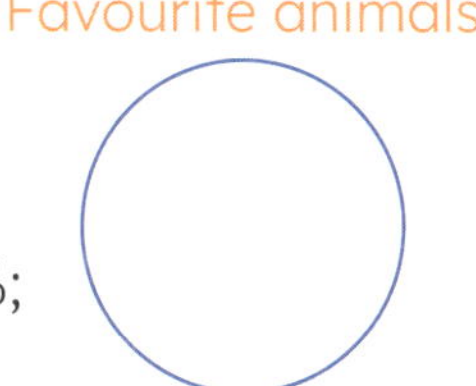

9. 0.79 × ______ = 790

10. 10 – 0.001 = ______

11. $\frac{1}{2} > \frac{1}{3}$ true ☐ false ☐

12. $5\frac{2}{3} > 5\frac{3}{5}$ true ☐ false ☐

13. $50.00 – $18.50 = ______

14. The prime factors of 4 are 2 × 2.

    The prime factors of 6 are ______

    and ______.

15. A, B, C = 30%, so D = ______%

    and E = ______%.

    What is the chance of an E outcome?

    (a) 0.2 ☐ (b) 0.4 ☐

    (c) 0.6 ☐ (d) 0.8 ☐

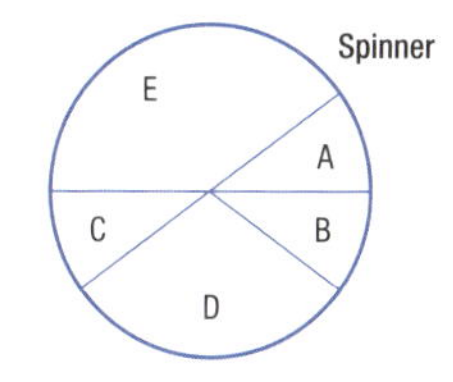

Week 13

Problem-solving

The population of Australia is almost 27 000 000 people (2023).

Rewrite this number in three different ways using exponent notation for powers of 10.

Read the question again. **Think** about the information. Underline the important words.

Tick the strategy you will use to work out the answer:

- estimate and check ☐
- look for patterns ☐
- draw a diagram or picture ☐
- construct a table or graph ☐
- use materials ☐
- use a formula ☐
- something else. ☐

Solve it:

Reflect on the question and answer.

Check it. Circle another strategy on the list to work it out.

Show it:

Friday Review

1. 1 000 000 =
   × ______ × ______
   × ______ × ______
   × ______ × ______
   $= 10^6$
2. Simplify $\frac{14}{20}$. ______
3. 0.038 × ______ = 3.8
4. Measure line $\overline{XY}$.
   ______ mm
   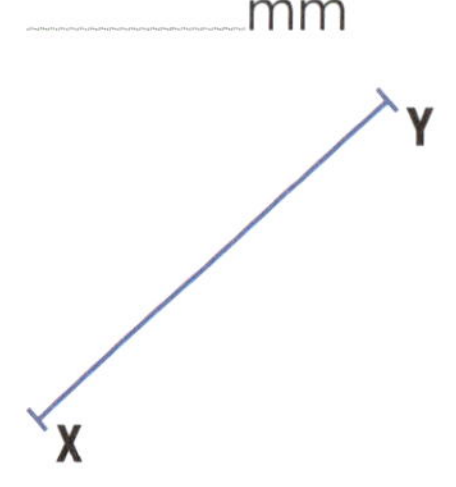

5. $9^2 = 6^2 + 3^2$
   true ☐ false ☐
6. 10 – 0.002 = ______
7. In Thursday's pie chart, are lemurs or bats more popular?
   ______
8. 5 + 0.006 + 0.2 + 0.09
   = ______
9. How many 20c coins make up $12.00?
   ______
10. $\frac{2}{3}$ = 0.______
11. What is the chance of you taking the bus home today?
    (a) certain ☐
    (b) unlikely ☐
    (c) very likely ☐
    (d) even ☐
12. If $\frac{3}{4} + y = 1$, then
    $y =$ ______.
13. Plot: A (3,4)
    B (1,4)
    C (4,3)
    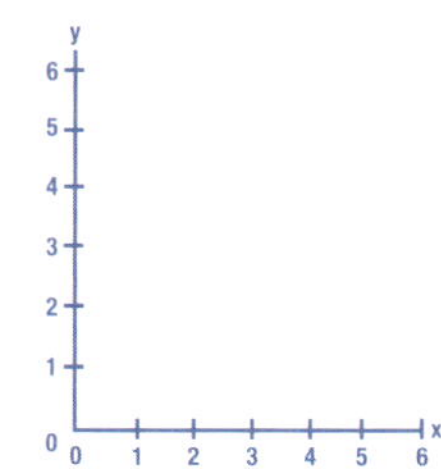

14. 2.02 < 2.1
    true ☐ false ☐
15. Round 4.072 to the nearest hundredth.
    ______
16. $3\frac{3}{8} > 3\frac{1}{2}$
    true ☐ false ☐
17. If ⬆ is north, what direction is B from A?
    ______
    B
    A
18. Draw the top view.

N A M Sp St P

Monday

1. Round 2.375 to the nearest tenth. ________

2. Write the time 25 minutes before.

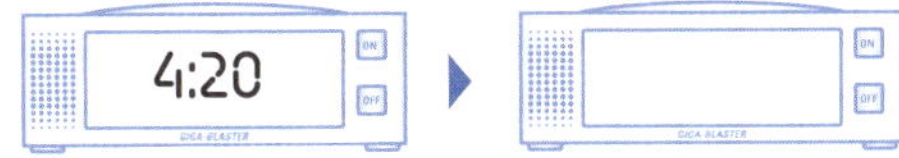

3. If $2\frac{1}{2} + y = 3$, then $y =$ ________.

4. Spin the cone on its axis 270° clockwise. Draw the new position.

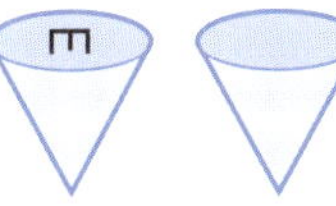

5. Which pair are divisible by 3?
   (a) 717, 293 ☐ (b) 525, 414 ☐
   (c) 452, 525 ☐ (d) 643, 172 ☐

6. $8 \times 10^4 =$ ________

7. What is the simplified ratio of blue cars to white cars if there are 100 blue and 300 white cars?

____

8. 95 + 65 = ________

9. $9\frac{2}{3} + 9\frac{2}{3} =$ ________

10. Match each volume amount to its jar.
    (a) 400mL (b) 410mL (c) 420mL

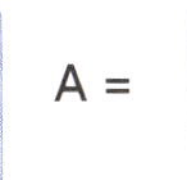

A = ________ B = ________ C = ________

A < B B > C C > A

11. Double $\frac{1}{4}$. ________

12. $10\% = \frac{10}{100} = 0.$________

13. If ⬆ is north, what direction is it from A to B?
    (a) NE ☐ (b) NW ☐
    (c) SE ☐ (d) SW ☐

14. A square has a perimeter of 24cm.
    What is its area? ________

15. The prime factors that make up 8 are:
    ________ × ________ × ________.

Tuesday

1. Round 5.408 to the nearest hundredth.
   ____

2. What number is halfway? ________

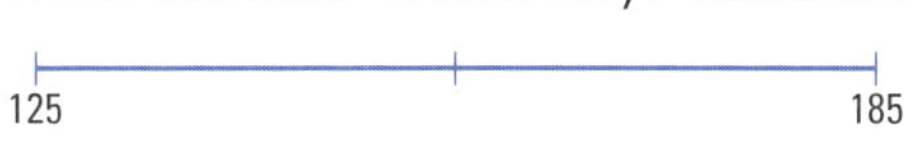

3. $5\frac{3}{8} =$ 5.38 ☐ 5.375 ☐ 5.25 ☐

4. $\frac{1}{2} < \frac{1}{5}$ true ☐ false ☐

5. What is the mean of 20, 30, 50, and 20?
   ____

6. What is the median number? ________

7. What is the mode? ________

8. Write the time 35 minutes after.

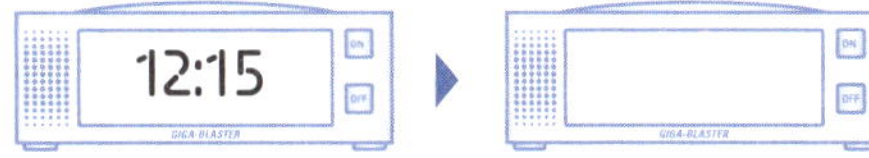

9. A fast train averages 150km/h for 4 hours.
   What distance does it cover? ________

10. 49.25 = 40 + 9 + ________ + ________

11. If ⬆ is north, what direction is it from B to A?
    (a) NE ☐ (b) NW ☐
    (c) SE ☐ (d) SW ☐

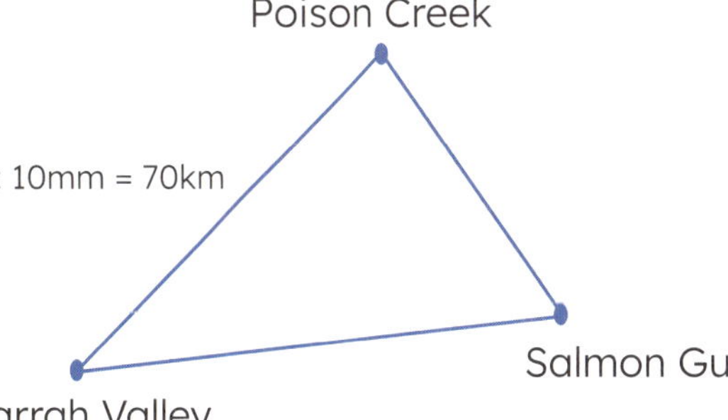

12. What is the distance from Poison Creek to Jarrah Valley? ________

13. What is the distance from Salmon Gully to Jarrah Valley? ________

14. Write in ascending order: $\frac{1}{2}, \frac{1}{3}, \frac{3}{4}, \frac{2}{5}$.
    ________, ________, ________, ________

15. If $6\frac{1}{2} + y = 7$, then $y =$ ________.

Week 14

Wednesday

1. Round 23.587 to the nearest hundredth.

______

2. Which decimal is closest to 4.25?

(a) 4.2 ☐ (b) 4.26 ☐
(c) 4.248 ☐ (d) 4.253 ☐

3. Draw another rectangle with the measurements reduced by one half.

(Not to scale.)

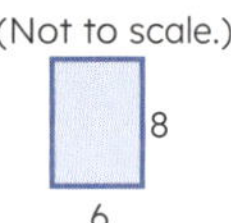

4. By how many times has the area of the rectangle been reduced?

(a) ÷ 4 ☐ (b) ÷ 1 ☐ (c) ÷ 8 ☐
(d) ÷ 6 ☐ (e) ÷ 12 ☐

5. Simplify $\frac{20}{24}$. ______

6. Looking at the cube's net, which face is opposite B?

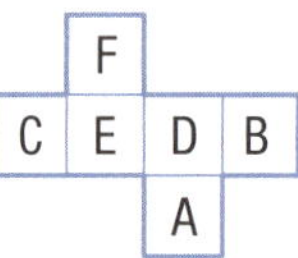

______

7. 1200m = (a) 1.2m ☐ (b) 1.20m ☐
(c) 1.2km ☐ (d) 12km ☐

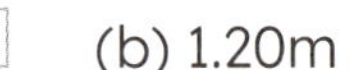

8. 147 + 69 = ______

9. If $y \times 10 = 5$, then y = ______.

10. 100% = $\frac{100}{100}$ = ______

11. –14 > –18 true ☐ false ☐

12. y = ______ °

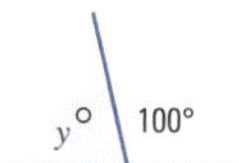

13. In a jar there are 10 marbles numbered from 1 to 10. What is the chance of you selecting a 4?

(a) 0.4 ☐ (b) 0.1 ☐ (c) 1 ☐
(d) 0.9 ☐ (e) 0.6 ☐

14. 2^4 = ______

15. A tap leaks at 0.25 litres per minute. How much water is wasted after 10 minutes?

______

Thursday

1. Round 2.9418 to the nearest hundredth.

______

2. 32 ÷ 4 + 4 × 2 = ______

3. If it is 2:00 am in Sydney (AEST), what time is it in Perth (AWST)?

______

4. 0.01 > 0.0009 true ☐ false ☐

5. What 2 prime numbers make up 10?

______, ______

6. 15% of $40.00 = ______

7. What 2 prime numbers add up to 28?

______, ______

8. Find the angle size.

a = ______

b = ______

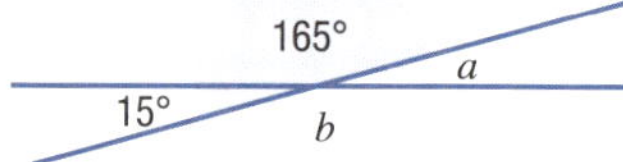

9. A chef measures $6\frac{2}{3}$ cups of flour. They use a $\frac{2}{3}$ measuring cup. How many $\frac{2}{3}$ cups does the chef measure?

______

10. If $n \div 3 = 6$, then n = ______

11. 0.04 = $\frac{4}{100}$ = ______%

12. Draw the front view.

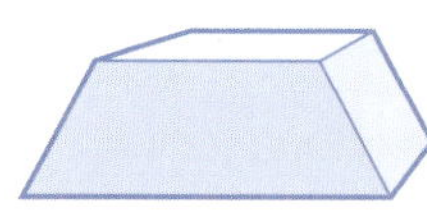

13. Share $50.00 equally among four people.

$______ each

14. 2^6 = ______

15. 5______ + 97 = 155

Problem-solving

Reggie plans to put skirting boards around the perimeter of the lounge room, which measures 14m by 6m. The room has three 80cm doorways around its edges, where skirting boards will not be needed. The skirting boards come in 3m lengths.

How many 3m lengths of skirting board will Reggie need to buy?

Read the question again. **Think** about the information. Underline the important words.

Tick the strategy you will use to work out the answer:

- estimate and check ☐
- look for patterns ☐
- draw a diagram or picture ☐
- construct a table or graph ☐
- use materials ☐
- use a formula ☐
- something else. ☐

Solve it:

Reflect on the question and answer.

Check it. Circle another strategy on the list to work it out.

Show it:

Friday Review

1. In Wednesday's marble jar (Q13), what is the chance of picking a 1?

   (a) 0.4 ☐ (b) 0.1 ☐

   (c) 0.9 ☐ (d) 0.6 ☐

2. Write the time 55 minutes before.

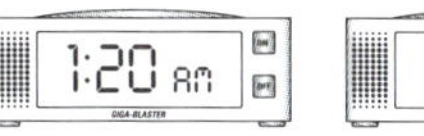

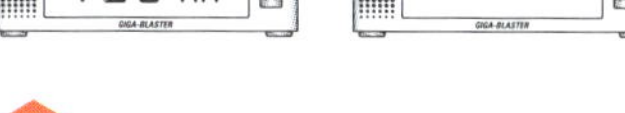

3. Draw a reflection.

(Not to scale.)

2cm

5cm

4. Double the size. Show the new measurements.

5. The area has increased by ________.

6. A tap leaks at 0.45L per minute. How much water is wasted after 5 minutes?

   ____

7. 250, 1000, 4000, 16 000, ________

8. 15% of $30.00 = ________

9. $42 \div n = 6$

   $n =$ ________

10. $2^5 =$ ________

11. Round 45.0968 to the nearest thousandth.

    ____

12. $y° =$ ________

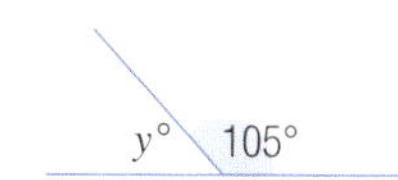

13. $6\frac{3}{4} + 3\frac{3}{4} =$ ________

14. Double $\frac{1}{4}$. ________

15. 6________ + 94 = 161

16. If you turn the original triangle by 270° clockwise, what is the new position?

    ____

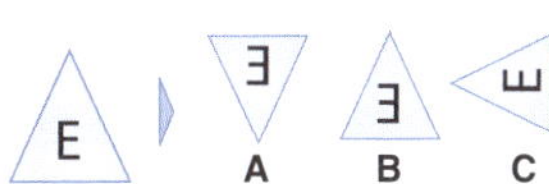

17. Match each volume to its jar.

    C > B   A < B

    380mL, 405mL, 0.41L

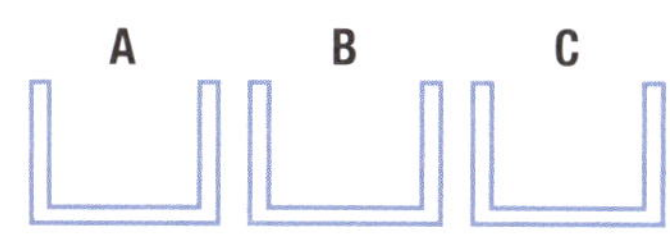

18. What is the mean of 50, 10, 20, and 20?

    ____

Monday

1. This shape has ______ pairs of parallel lines and ______ perpendicular lines.

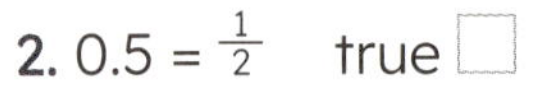

2. $0.5 = \frac{1}{2}$ true ☐ false ☐

3. Write 0.2 on the line.

4. If in Tarndanya / Adelaide it is 5:00 am (ACST), what is the time in Boorloo / Perth (AWST)?

______

5. What is the value of 2 in 7.002?

______

6. If you turn the rectangle 270° clockwise, what will the new position be?

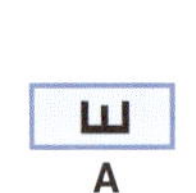

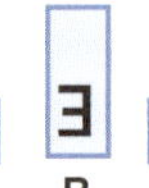

______

7. What is the probability of randomly selecting a king, queen, or jack from a deck of 52 cards?

______

8. $-15 < -23$ true ☐ false ☐

9. Round 5.408 to the nearest hundredth.

______

10. $3\frac{7}{10} + 2\frac{5}{10} =$ ______

11. $\frac{3}{100} + \frac{7}{10} + \frac{4}{100} =$ ______

12. 1, 1.8, 2.6, 3.4, ______

13. Draw the right side view.

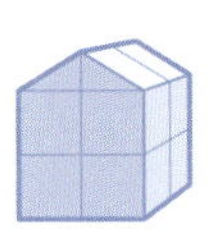

14. $5\overline{)800} =$ ______

15. Which expression matches? There are five bananas in each 1kg bag. How many bags were needed for 200 bananas?

(a) 200×5 ☐ (b) $200 \div 5$ ☐

(c) 5×200 ☐ (d) $200 + 5$ ☐

Tuesday

1. This shape has ______ pairs of parallel lines and ______ perpendicular lines.

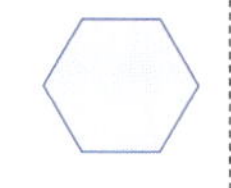

2. $3\frac{5}{8} < \frac{15}{4}$ true ☐ false ☐

3. If $2 - y = 1\frac{1}{3}$, then $y =$ ______.

4. $0.01 \neq \frac{1}{10}$ true ☐ false ☐

5. 25% of $40.00 = ______

6. $2.75 \div$ ______ $= 0.275$

7. Record 0.05 on the line.

8. $10^6 =$ ______

9. 5kL = ______L

10. If $n \times 6 = 42$, then $n =$ ______.

11. $56 \div 7 + 4 \times 5 =$ ______

12. Arlo (a) scored 8 more runs than Kieran (k) during a cricket game. Which expression matches the number of runs Arlo scored?

(a) $a + 8$ ☐ (b) $a - 8$ ☐

(c) $k - 8$ ☐ (d) $k + 8$ ☐

13. Round 8.069 to the nearest hundredth.

______

14. Which expression matches? A blueberry farm packs 100g of fruit in each punnet. If they pack 50 punnets, what is the total mass in grams?

(a) 100×50 ☐ (b) $100 \div 5$ ☐

(c) $50 \div 100$ ☐ (d) $100 + 50$ ☐

15. $8.3 - 0.5 =$ ______

Wednesday

1. y = ________ ° ($y°$, 45°)

2. Round 2.349 to the nearest hundredth.

____

3. 1 000 000 – 10 = ________

4. After constructing this cube, which letter is opposite F?

A
B C D E
F

____

5. The LCD for $\frac{1}{4}$ and $\frac{3}{10}$ is ________.

6. Halve 4.9. ________

7. If it is midday in Melbourne (AEST), what time is it in Alice Springs?

____

8. If ↑ is north, what direction is X to Y?

(a) NE ☐ (b) NW ☐
(c) SE ☐ (d) SW ☐

9. 12 × 8 = 24 × ________ = ________

10. 4.07 ÷ ________ = 0.0407

11. Turn the rectangle 180°. Draw the new position.

12. Share $100 equally among 8 people.

$________ each

13. Shade the rectangle to show the percentages of types of lunch.

Lunch survey:

Eggs	(E)	20%
Pies	(P)	40%
Nachos	(N)	10%
Pizza	(Pi)	30%

14. $\frac{29}{100}$ = ________%

15. If apples cost $4.98 per kg and a box contains 8kg, what is the approximate cost of four boxes?

(a) $32 ☐ (b) $40 ☐
(c) $160 ☐ (d) $128 ☐

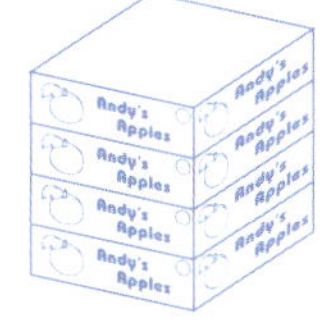

Thursday

1. a = ________ °

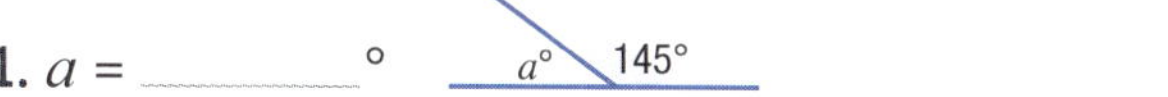

2. Shade the side that is lighter.

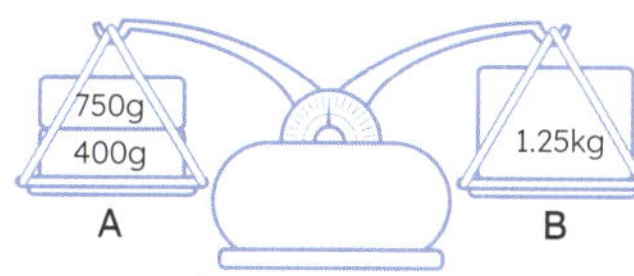

3. Double 4.9. ________

4. 100 = ________ × ________ = 10^2

5. The prime factors of 12 are ________.

6. 80 000 + 130 000 = ________

7. Rounding π to 3, what is the area of a circle if its radius is 5cm?

____

8. Record 0.05 on the line.

9. 4 + 9 × 2 = ________

10. Circle the prime numbers.

11, 12, 13, 14, 15, 16, 17, 18, 19

11. A car is travelling at 100km/hr. How far will it travel after $\frac{3}{4}$ of an hour?

________km

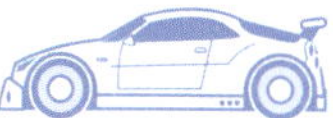

12. 25% of $100.00 = ________

13. If $9\frac{1}{2} + y = 10$, then y = ________.

14. Matt exchanged A$50 for US$40. How many USD could he get with A$200?

____

15. $8\frac{1}{4}$ = ____.____

Week 15

Problem-solving

What are the sizes of angles a, b, c, d, e, and f?

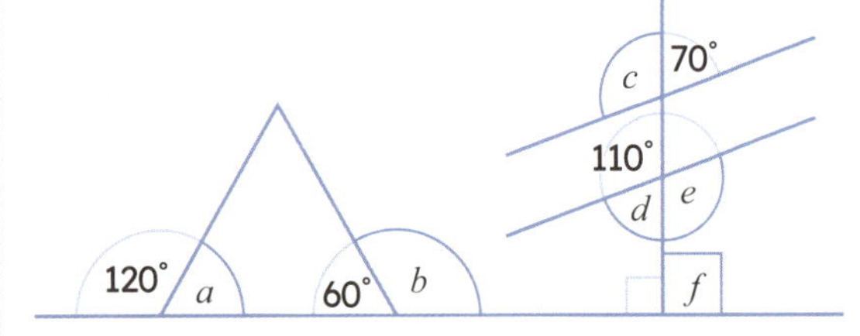

Read the question again. **Think** about the information. Underline the important words.

Tick the strategy you will use to work out the answer:

- estimate and check ☐
- look for patterns ☐
- draw a diagram or picture ☐
- construct a table or graph ☐
- use materials ☐
- use a formula ☐
- something else. ☐

Solve it:

Reflect on the question and answer.

Check it. Circle another strategy on the list to work it out.

Show it:

Friday Review

1. Round 5.408 to the nearest hundredth. ______

2. If diameter = 10cm, then radius = ______ cm.

3. Matt exchanged A\$50 for US\$40. How many USD could Matt get with A\$400? ______

4. 1000 = ______ × ______ × ______ = 10^3

5. $\frac{36}{10} + \frac{27}{10} =$ ______

6. 25% of \$40.00 = ______

7. Which is the heavier side? ______

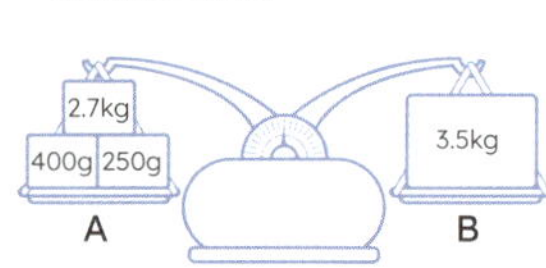

(Not to scale.)

8. In Gulumoerrgin / Darwin (ACST) the time is 1:00 am. What is the time in Boorloo / Perth (AWST)? ______

9. Which expression matches?

   Beans are packed in 500g bags. How many bags are needed for 5kg?

   (a) 5 ÷ 500 ☐
   (b) 5000 ÷ 500 ☐
   (c) 500 ÷ 5 ☐

10. 2.07 ÷ ______ = 0.027

11. 8kL = ______ L

12. $y° =$ ______

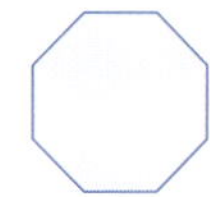

13. This shape has ______ pairs of parallel lines and ______ perpendicular lines.

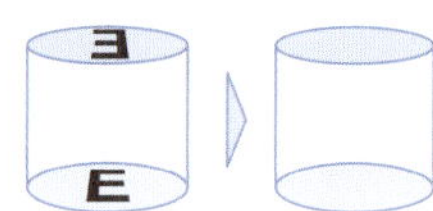

14. 15% = 0.______

15. Rotate the cylinder 180° clockwise. Draw the new position.

16. $0.01 \neq \frac{1}{100}$

    true ☐ false ☐

17. 49 ÷ 7 + 7 × 3 = ______

18. If a car is travelling at 100km/h, how far does it travel in 15 minutes? ______

Monday

1. Area of triangle

= ________ cm^2

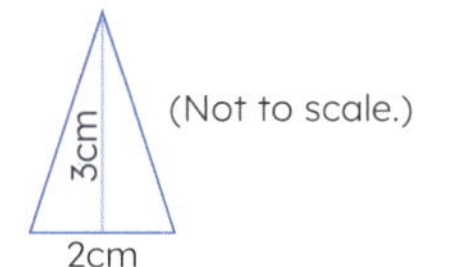

2. If 12 500 – 6500 = a, then a = ________.

3. 9.3 – 0.9 = ________, 0.93 – 0.09 = ________

4. After five games, Natasha scored 14, 8, 12, 6, and 2 for the netball team.
   What is the average (mean)? ________

5. The LCD for $\frac{1}{4}$ and $\frac{3}{5}$ is ________.

6. Which two prime numbers are the sum of 12?

   ________ ________

7. Turn the rectangle 450° clockwise. Draw the new position.

8. If $y \times 3 = \frac{3}{5}$, then y = ________.

9. Double 17 585. ________

10. $3\frac{4}{5}$ = ________.________

11. How many 50c coins make up $50? ________

12. Order the competitors from first to last. Alicia was ahead of Linny by five minutes. Kate was faster than Linny by two minutes. Sonia was ahead of Alicia by five minutes.

    1st ________

    2nd ________

    3rd ________

    4th ________

13. $5 \times 10^5 + 3^2$ = ________

14. What is the ratio of Sydney United football fans to Brisbane County fans if there are 1000 Sydney fans and 200 Brisbane fans?

    ________ in ________

15. If it is midnight in Naarm / Melbourne (AEST), what is the time in Boorloo / Perth?

    ____

Tuesday

1. Area of triangle

= ________ cm^2

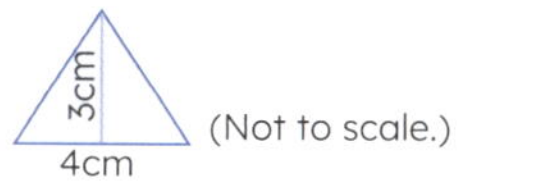

2. 16 = 10 + ________ × ________

3. 2.3 + 0.8 = ________

4. 4 × 7 = ________

5. In a pie chart, graph the data about favourite football teams: Bendigo Bats (BB) 5%; Perth Panthers (PP) 25%; Adelaide Adders (AA) 30%; Darwin Dingoes (DD) 30%; Hyden Hammers (HH) 10%.

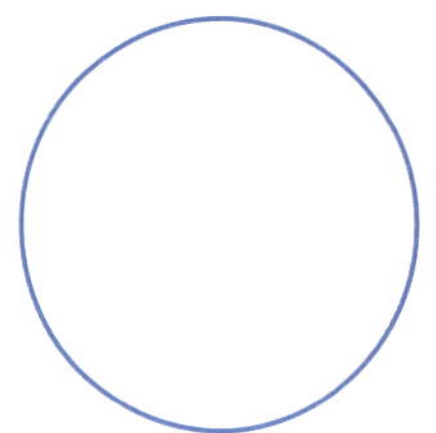

6. 25.5 ÷ 100 = ________

7. The LCD for $\frac{2}{3}$ and $\frac{1}{6}$ is ________.

8. 6^2 = (a) 4 × 9 ☐ (b) 5 × 6 ☐ (c) 2 × 6 ☐ (d) 6 + 6 ☐

9. a = ________ °

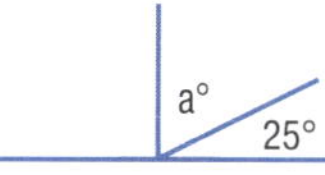

10. 70 000 + 340 000 = ________

11. If you enlarge a rectangle by doubling the length of each side, by how many times will its area increase?

    ____

12. Double 9995. ________

13. Which expression matches? A farrier has 48 horseshoes. How many horses can they fit shoes to?

    (a) 48 ÷ 4 ☐ (b) 48 + 4 ☐
    (c) 48 – 4 ☐ (d) 4 + 48 – 4 ☐
    (e) 48 – 0 ☐

14. Round π to the nearest whole. ________

15. 2^5 = ________

Week 16

Week 16

Wednesday

1. Area of triangle = ________ cm^2

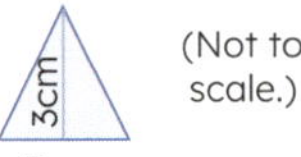

(Not to scale.)

2. Simplify $\frac{14}{18}$. ________

3. 66 – 8 = ________

4. To share $25 with 20 teachers:

   $25 ÷ 20 = $25/20 = $ ________ each

5. $(3 \times 10^5) + (4 \times 10^4) + (8 \times 10^3) + (9 \times 10^2)$

   = ________

6. 10 – 0.004 = ________

7. 10kL = ________ L

8. If $8 = y + 6\frac{1}{2}$, then y = ________.

9. What is the probability of choosing a red marble from a jar if there are 15 blue, 10 red, and 25 green marbles?

____

10. $\frac{2}{5} + \frac{1}{2} = \frac{\square}{100}$

11. Which two towns are 218km apart?

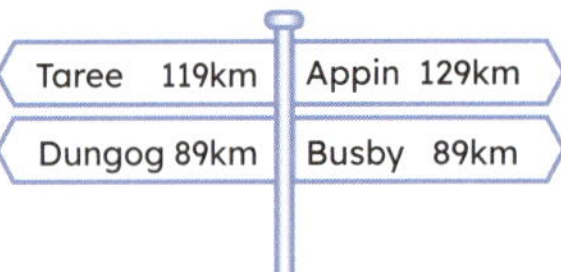

____

____

12. Draw arrows to show the direction of the path between the two points.

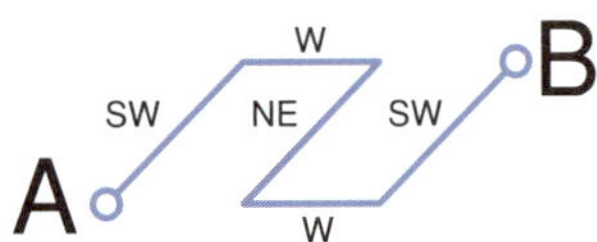

13. What is the area of this floor? ________ m^2

(Not to scale.)

14. $0.009 \neq \frac{9}{100}$ true ☐ false ☐

15. 60% of $110.00 = ________

Thursday

1. Area of triangle = ________ cm^2

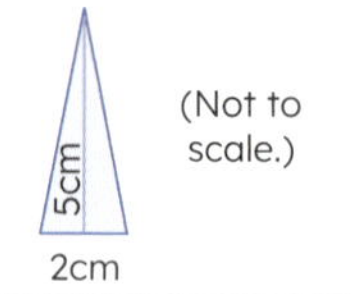

(Not to scale.)

2. Alex was 13 minutes faster in the triathlon race than Jevan. In which position did Jevan finish?

   1st: 1 hr 58 min.

   2nd: 2 hr 8 min.

   3rd: 2 hr 11 min.

   ____

3. A chef has a $\frac{2}{3}$ measuring cup and measures out nine cupfuls of sugar. How many whole cups does this equal?

   ____

4. Share $90 equally among 6 people.

   ________, ________, ________, ________, ________, ________

5. 11 × 10 = ________

6. 1.22 = ________ %

7. Continue the pattern.

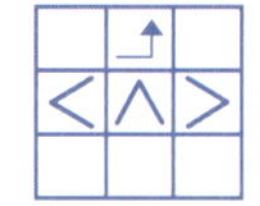
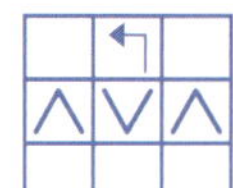
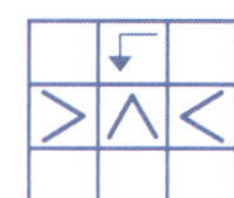
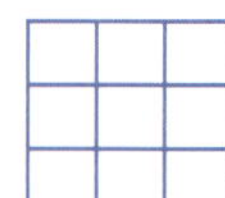

8. 1 000 001 – 8 = ________

9. $6\frac{3}{4} + 3\frac{3}{4}$ = ________

10. If $y \times 10 = 50$, then y = ________.

11. 800 + 500 = ________

12. 2 + 9 × 3 = ________

13. Write 9:15 pm in 24-hour time. ________

14. $50.00 – $11.25 = ________

15. A red car is travelling at 90km/hr. How far will it travel in 20 minutes?

   ____

Problem-solving

Rianna and Birrani were making bunting to hang in their garden. They had exactly enough material for 15 isosceles triangles (base 20cm and height 30cm) and 14 rectangles (length 35cm and width 20cm).

What was the area of the material used to create all the triangular and rectangular bunting flags?

Read the question again. **Think** about the information. Underline the important words.

Tick the strategy you will use to work out the answer:

- estimate and check ☐
- look for patterns ☐
- draw a diagram or picture ☐
- construct a table or graph ☐
- use materials ☐
- use a formula ☐
- something else. ☐

Solve it:

Reflect on the question and answer.

Check it. Circle another strategy on the list to work it out.

Show it:

Friday Review

1. In Wednesday's jar, what is the probability of choosing a green marble? ______
2. \$50.00 – \$13.05 = ______
3. $7\frac{3}{4} + 2\frac{3}{4}$ = ______
4. Complete the pattern.

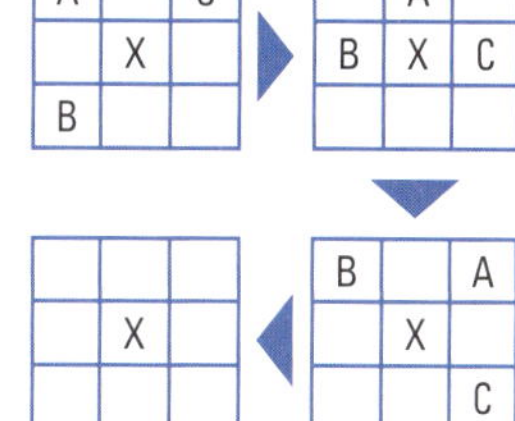

5. The LCD for $\frac{1}{2}$ and $\frac{1}{3}$ is ______.
6. Double 19 585. ______
7. If you enlarge a rectangle by doubling the length of each side, by how many times will its area increase? ______
8. Area of a triangle = ______ cm^2

(Not to scale.) 2cm, 5cm

9. 70% of \$90.00? = ______
10. If $y + 7\frac{1}{2} = 8$, then y = ______.
11. $6 \times 10^4 + 4^2$ = ______
12. $7\frac{1}{5}$ = ______.______
13. What is the mean of 6, 8, 16, and 10? ______
14. Draw the arrow a $\frac{3}{4}$ turn clockwise in the box.

15. $6\overline{)1206}$ = ______
16. $a°$ = ______

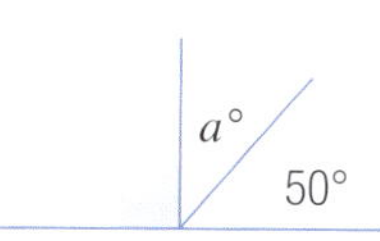

17. Round 0.153 to the nearest hundredth. ______
18. If there are 2000 West Coast Pigeons' fans and 200 Sydney Ducks' fans, what is the ratio of Pigeon to Duck fans? ______

N A   M  Sp St P 

Week 17

Monday

1. Flip a coin. The two outcomes are

_______ or _______.

2. The LCD for $\frac{3}{4}$ and $\frac{2}{3}$ is _______.

3. $(5 \times 10^5) + (9 \times 10^3) + (7 \times 10^2) =$ _______

4. 80 000 + 53 000 + 27 000 = _______

5. Which fraction is between 0.4 and 0.67?

(a) $\frac{1}{4}$ ☐ (b) $\frac{1}{2}$ ☐

(c) $\frac{1}{8}$ ☐ (d) $\frac{1}{5}$ ☐

6. 28.7 ÷ 10 = _______

7. 9.6 – 0.9 = _______

8. What are the prime factors of 14?

____

9. $1\frac{35}{100}$ = 1.35 = _______%

10. If each month of the year is written on cards, what is the probability of randomly choosing a month beginning with J from a hat?

_______ in _______

11. Write one million and eleven as a numeral.

____

12. Halve 0.1. _______

13. What number is halfway? _______

80 — 110

14. If a car takes 10 minutes to travel 10km, how fast is it travelling?

____

15. $\frac{3}{1000}$ = 0._______

$\frac{11}{1000}$ = 0._______

Tuesday

1. Flip two coins. The four outcomes are HH,

_______, _______, or _______.

2. Measure:

$\overline{AB}$ = _______mm, $\overline{BC}$ = _______mm

A — B — C

3. Write $3\frac{1}{7}$ as a mixed number. _______

4. 6500 + 19 500 + 2500 = _______

5. Double 0.07. _______

6. $\frac{3}{4}$ of 16 = _______

7. 3 × 4 × 5 = 6 × _______ × 5

8. 2.02 = _______%

9. 270 000mm + 50 000mm = _______m

10. 10 + 9 × 2 + 14 ÷ 2 = _______

11. If a clock shows 3 o'clock, what is the size of the smaller angle between the hands?

____

12. If you halve the measurements of this shape, by what amount is the area reduced?

8cm, 2cm (Not to scale.)

(a) ÷ 1 ☐ (b) ÷ 2 ☐

(c) ÷ 3 ☐ (d) ÷ 4 ☐

13. If $n \div 8 = 2.5$, then $n =$ _______.

14. $\frac{1}{8}$ = 0._______

15. 25% of 120 = _______

Wednesday

1. Throw a die. The six outcomes are

______, ______, ______,

______, ______, ______.

2. The following are shoe sizes of basketball players: 12, 14, 13, 14, 13, 15, 17.

What is the mean? ______

3. Share $15 equally among four people.

$______ each

4. Shade 40mL for each container.

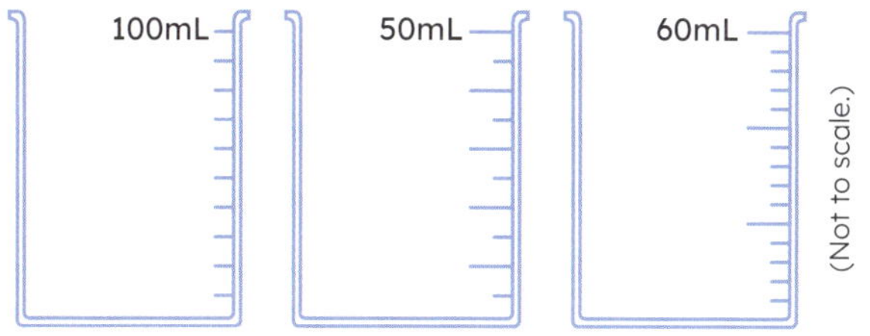

5. Which shoes are cheaper to buy?

(a) 40% off $120 ☐

(b) 20% off $100 ☐

(c) 10% off $90 ☐

6. Simplify $\frac{20}{30}$. ______

7. If $y \times 50 = 25$, then $y =$ ______.

8. $8 - \frac{3}{4} =$ ______

9. What is the angle size of:

$a°$? ______ $b°$? ______

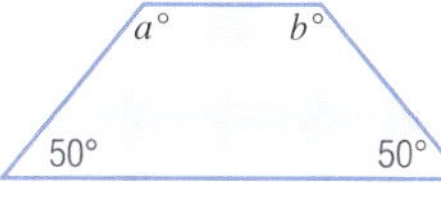

10. $10 \times$ ______ $= 10^4$

11. 5000mm = ______m

12. $a° =$ ______

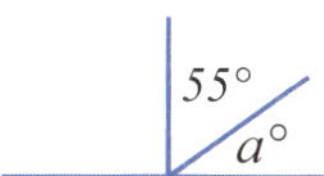

13. 0.97, 0.98, 0.99, ______

14. Write one hundred and eleven thousand and ten as a numeral.

______

15. If a square sheet of paper has a perimeter of 28cm, what is its area?

______cm²

Thursday

Week 17

1. A bag contains a red ball, a blue ball, and a pink ball. The three outcomes are

______, ______, or ______.

2. Draw the top view.

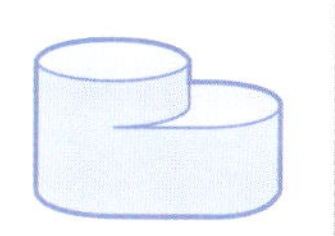

3. If it is 1:00 pm at the Sydney Opera House, what time will it be at the Melbourne Cricket Ground?

______

4. $9\frac{3}{4} + \frac{1}{2} =$ ______

5. Your money jar has: 17 × 20c, 11 × 10c, and 7 × 50c. What is the total amount?

______

6. If $n \div 6 = 5.5$, then $n =$ ______

7. $9^2 =$ ______

8. Which shape is:

(a) an octagon? ______

(b) a hexagon? ______

(c) a pentagon? ______

9. 999 999 + 11 = ______

10. 7 × 8 = ______

11. If the car was originally $3000, what is its price now?

______

12. If your new car travels 8km in 10 minutes, how fast is it going? ______km/h

13. 65 000 + 95 000 + 45 000 = ______

14. A box contains twice as many blue pens as red pens. In total there are 18 pens. How many blue pens are there?

______

15. How many B boxes will fit into box A?

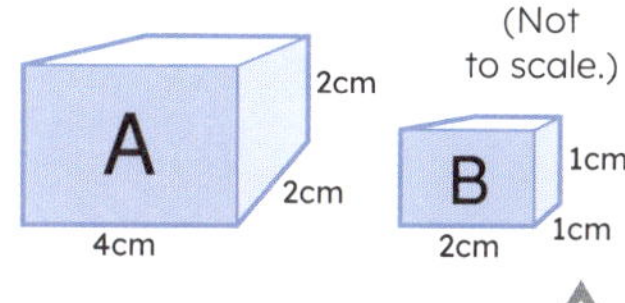

______

Problem-solving

Milo was playing with two spinners. Spinner A was divided into four equal segments and spinner B was divided into 12 equal segments. Spinner A had a 50% chance of landing on red whilst spinner B had half this chance of landing on red. Both spinners had a 25% chance of landing on blue. Spinner B had a $\frac{1}{3}$ chance of landing on yellow whilst spinner A had zero chance. The remaining segments of both spinners were green.

What chance, in fractions, did the spinners have of landing on green?

Read the question again. **Think** about the information. Underline the important words.

Tick the strategy you will use to work out the answer:

- estimate and check ☐
- look for patterns ☐
- draw a diagram or picture ☐
- construct a table or graph ☐
- use materials ☐
- use a formula ☐
- something else. ☐

Solve it:

Reflect on the question and answer.

Check it. Circle another strategy on the list to work it out.

Show it:

Friday Review

1. Flip two coins. The four outcomes are ________, ________, ________, or ________.
2. If y + $3\frac{3}{7}$ = 4, then y = ________.
3. 3.04 = ________%
4. Halve 0.3. ________
5. 25% of $200.00 = ________
6. Which fraction is between 0.2 and 0.4?
   (a) $\frac{1}{3}$ ☐ (b) $\frac{1}{8}$ ☐
   (c) $\frac{3}{5}$ ☐ (d) $\frac{3}{6}$ ☐
7. If it is 6:00 pm in Tarndanya / Adelaide, what time is it in Warrane / Sydney? ________
8. Measure line $\overline{XY}$. ________mm
9. What number is halfway between 70 and 140? ________
10. Which is a net of a cube? ________

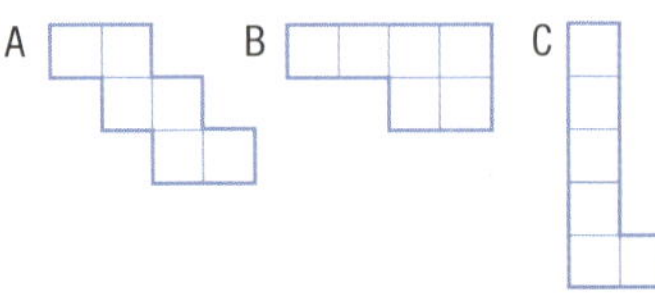

11. What is the mean of 15, 35, and 25? ________
12. 9000mm = ________m
13. 60.5 ÷ 100 = ________
14. Shade 25mL.

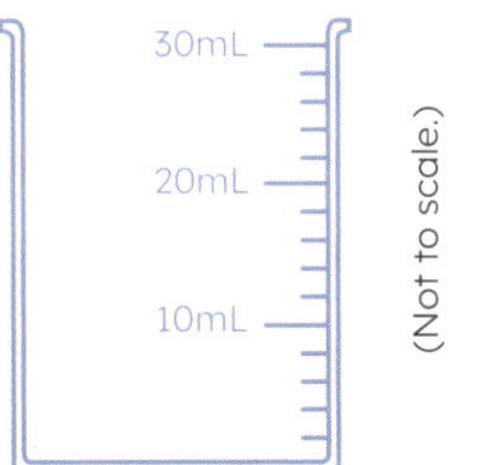

15. A bag contains a red ball, a blue ball, and a pink ball. You pick out two balls. The outcomes are ________, ________, ________, ________, ________, ________.
16. If n ÷ 6 = 2.5, then n = ________.
17. If a car takes 20 minutes to travel 15km, how fast is it going? ________
18. How many B boxes will fit into box A? ________

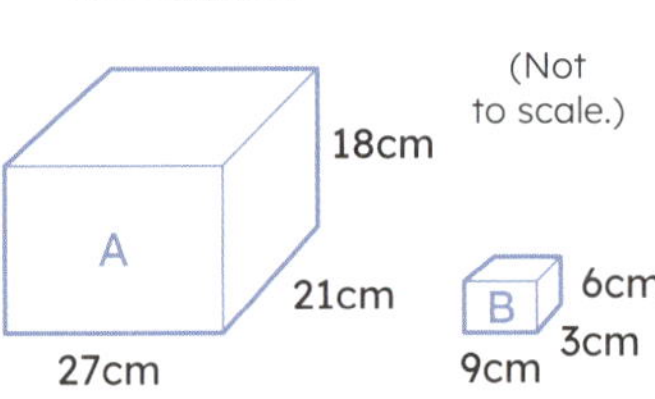

Monday

1. The formula for the area of a rectangle is

_______ × _______.

2. What is the time difference? _______

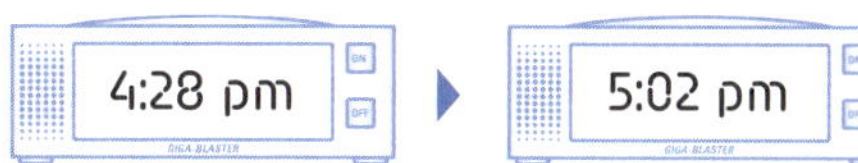

3. 45 000 + 255 000 = _______

4. $\frac{3}{5}$ of 125 = _______

5. 9.09 < 9.9 true ☐ false ☐

6. 30% of $40.00 is _______.

7. If there is $10.40 in 20c coins, how many coins are there?

____

8. If $n \div 4 = 12.25$, then $n =$ _______.

9. $8\frac{3}{4} + \frac{1}{2} =$ _______

10. How many days are there in August?

____

11. $(6 \times 10^6) + (2 \times 10^5) + (7 \times 10^4) +$

$(4 \times 10^3) + (5 \times 10^2) + (8 \times 1) =$ _______

12. If $7.2 - y = 6.8$, then $y =$ _______.

13. 27 − 8 = _______, 2.7 − 0.8 = _______

14. Double $\frac{3}{5}$. _______

15. If each mark is equivalent to 0.5mL, what is the level at A?

____

(Not to scale.)

Tuesday

1. The formula for the area of a square is $a^{\square}$.

2. $\frac{2}{3}$ of 36 = _______

3. What are the prime factors of 18?

____

4. The LCD for $\frac{3}{4}$ and $\frac{1}{6}$ is _______.

5. Write ten million, one hundred and ten thousand as a numeral.

____

6. 1000 =

_______ × _______ × _______ $= 10^3$

7. If a car travels 12km in 10 minutes how fast is it going?

_______km/h

8. What is the probability of randomly selecting an even number from a pack of 52 cards?

____

9. Area of triangle = _______cm^2

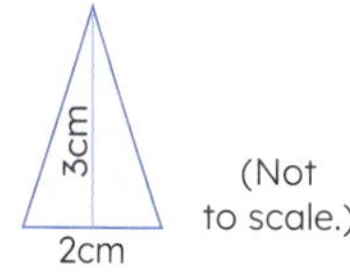

(Not to scale.)

10. Share $500.00 equally among 20 people.

$_______ each

11. What number is halfway between? _______

120 ———————— 190

12. $8 \times \frac{1}{5} =$ _______

13. 0.5 × 75 = _______

14. Halve $\frac{4}{10}$. _______

15. 1, 3, 9, 27, _______, 243

Week 18

Week 18

Wednesday

1. The formula for the area of a circle is $\pi r^{\square}$.

2. The LCD for $\frac{2}{3}$ and $\frac{1}{5}$ is ________.

3. 95 000 + 125 000 = ________

4. Simplify $\frac{21}{28}$. ________

5. 1 000 000 – 100 000 = ________

6. $a°$ = ________ °

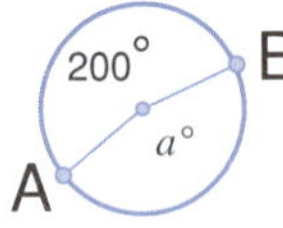

7. 4.2m = ________ mm

8. $\frac{4}{5} > \frac{2}{10}$ true ☐ false ☐

9. An icosahedron has ________ faces.

10. $x°$ = ________ °

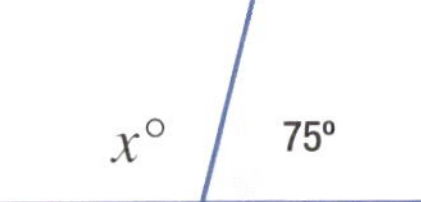

11. If 4.0 + y = 7.5, then y = ________.

12. A dodecahedron has ________ faces.

13. Draw the shape's lines of symmetry.

How many are there? ________

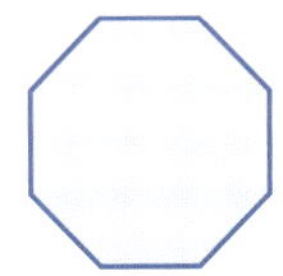

14. What two prime numbers add up to 30?

________, ________

15. 443 + 57 = ________

Thursday

1. The formula for the area of a triangle is

$\frac{1}{2}$ ________ × ________.

2. What is the time difference? ________

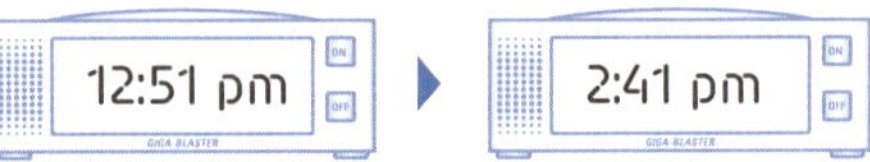

3. If there are 2000 Sydney fans and 500 Brisbane fans, what is the ratio of Sydney to Brisbane fans?

____

4. 1 = ________ %

5. A bricklayer laid 145 bricks in two hours. They laid 25 bricks more in the first hour. How many bricks did they lay in the second hour?

____

6. $\frac{4}{5}$ of 80 = ________

7. $4\frac{1}{2} + \frac{3}{4}$ = ________

8. What is the probability of randomly selecting an odd number card from a pack of 52 playing cards?

____

9. $d°$ = ________ °

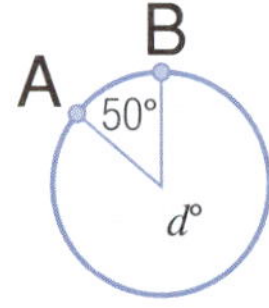

10. $\frac{7}{100}$ = 0.07, $\frac{7}{1000}$ = ________, $\frac{70}{1000}$ = ________

11. 10m = ________ mm

12. If a chef used a $\frac{3}{4}$ measuring cup to measure out 12 cups of flour, how many $\frac{3}{4}$ cups did they use?

____

13. 6 × 4 = 3 × ________ = ________

14. If n + 18 = 37, then n = ________

15. 10^2 = ________

Problem-solving

Alinta needs to buy a new sail for this yacht.

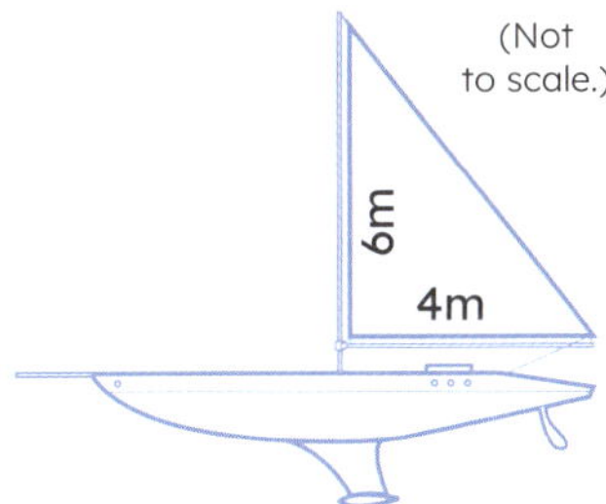

If a new sail costs $55 per m^2, how much will Alinta need to spend?

Read the question again. **Think** about the information. Underline the important words.

Tick the strategy you will use to work out the answer:

- estimate and check ☐
- look for patterns ☐
- draw a diagram or picture ☐
- construct a table or graph ☐
- use materials ☐
- use a formula ☐
- something else. ☐

Solve it:

Reflect on the question and answer.

Check it. Circle another strategy on the list to work it out.

Show it:

Friday Review

1 The formula for the area of a rectangle is ______ × ______.

2 What is the probability of randomly selecting an even number card from a 52-card deck?

______

3 Round 0.639 to the nearest tenth.

______

4 What is the volume of a room which measures 10m × 7m × 3m?

______

5 65 000 + 325 000 = ______

6 100 000 = ______ × ______ × ______ × ______ × ______ = 10^5

7 If 6.2 + y = 7.9, then y = ______.

8 The formula for the area of a circle is $\pi r^{\square}$.

9 1, 4, 16, 64, ______, 1024

10 $7\frac{9}{10} + \frac{3}{10}$ = ______

11 What number is halfway between 140 and 210?

______

140 ——|—— 210

12 Which two prime numbers add up to 20?

______, ______

13 An icosahedron has ______ faces.

14 $\frac{8}{10}$ = 0.______, $\frac{8}{1000}$ = 0.______, $\frac{80}{1000}$ = 0.______

15 $a°$ = ______

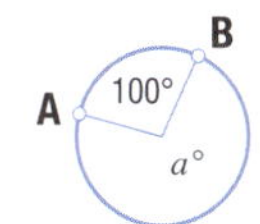

16 Area of triangle = ______cm^2

(Not to scale.)
5cm
2cm

17 If there is $12.60 in 20c coins, how many coins are there?

______

18 $\frac{4}{5}$ of 120 = ______

Sp St P

Week 19

Monday

1. Complete the pattern.

2		1
	3	

	5	
10		
		15

		375
125		
	250	

2. Two coins were flipped one after another. Write the four possible outcomes.

_______, _______, _______, _______

T = tails H = heads

3. 3 × 7 × 4 = _______

4. A car travels constantly at 100km/h for 350km. How many hours is the trip?

____

5. 200 – 50 ÷ (40 ÷ 8) = _______

6. 1 000 000 – 50 000 = _______

7. $c°$ = _______ °

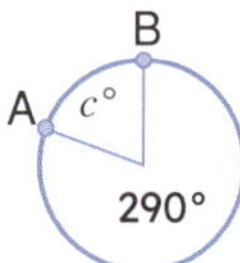

8. $6\frac{1}{3} + 2\frac{2}{3}$ = _______

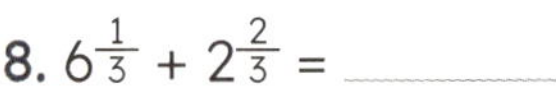

9. Hannah the hairdresser cuts hair for three days each week. In total, Hannah cuts 123 heads of hair. What is the average number of clients each day?

____

10. 35% of $1800 = _______

11. Which two are not nets for a cube? _______

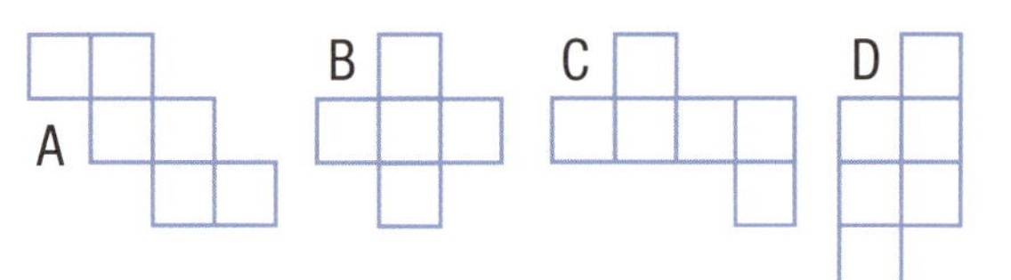

12. $9\frac{2}{3} + 8\frac{5}{6}$ = _______

13. If 3 – y = 2.7, then y = _______.

14. Add 75 minutes to the time.

15. If a bus leaves its depot at 4:15 pm and has to stop at 5 locations at 15 minute intervals, what time is it by the third stop?

____

Tuesday

1. If each number is a multiple, what number is C?

____

36	B	C
D	24	F
G	H	6

36	B	C
48	E	18
G	H	I

A	12	C
D	24	F
G	30	I

A	12	C
D	E	F
42	H	6

2. Add six and one-quarter hours to the time.

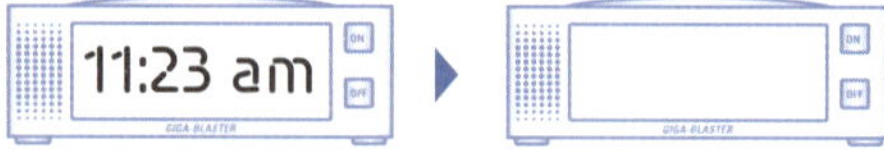

3. A rectangular-shaped warehouse has a perimeter of 80m. If its length is 25m, what is its width?

____

4. Draw all the shape's lines of symmetry.

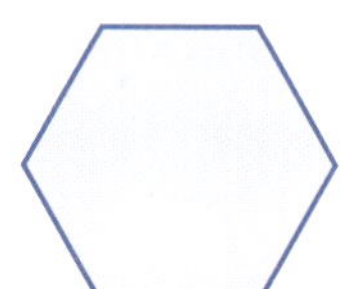

5. (600 – 200) ÷ (10 ÷ 2) = _______

6. The LCD for $\frac{3}{10}$ and $\frac{1}{6}$ is _______.

7. A football match at Octopus Oval in Warrane / Sydney begins at 12:30 pm (AEST). What is the starting time in Boorloo / Perth?

____

8. $a°$ = _______ ° 135° $a°$

9. If 9 – y = 8.5, then y = _______.

10. (a) 72 – 8 = _______

(b) 720 000 – 80 000 = _______

11. Draw the front view.

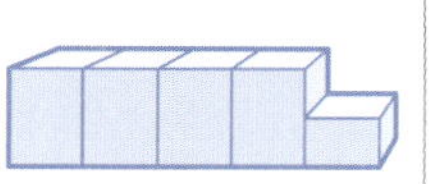

12. Shade 0.6mL

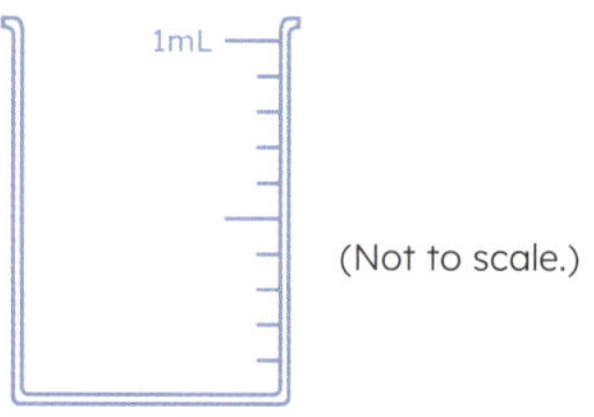

(Not to scale.)

13. If 32 = 2a, then a = _______.

14. Write the factors of 48.

____

15. How many days are in October, November, and December combined?

____

Wednesday

1. Continue the pattern.

		Y
		△
	A	X

	Y	△
		X
		A

Y	△	X
		A

2. What two prime numbers equal the sum of 22?

_______, _______

3. Draw the reflection.

4. The LCD for $\frac{2}{10}$ and $\frac{3}{4}$ is _______.

5. A rectangular-shaped warehouse has a perimeter of 70m. The width is 15m.

The length is _______m.

6. Write one million, nine hundred and nine thousand as a numeral.

____

7. If $y \times 12 = 24$, then $y =$ _______.

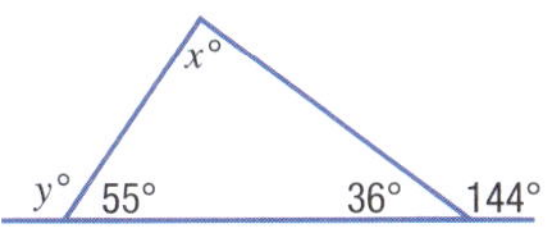

8. Without using a protractor, what is the size of angle x? _______

9. What is the size of angle y? _______

10. Does 2117 + 3119 + 4246 equal an odd or even number?

odd ☐ even ☐

11. Write the new measurements of each side if all the side lengths were doubled.

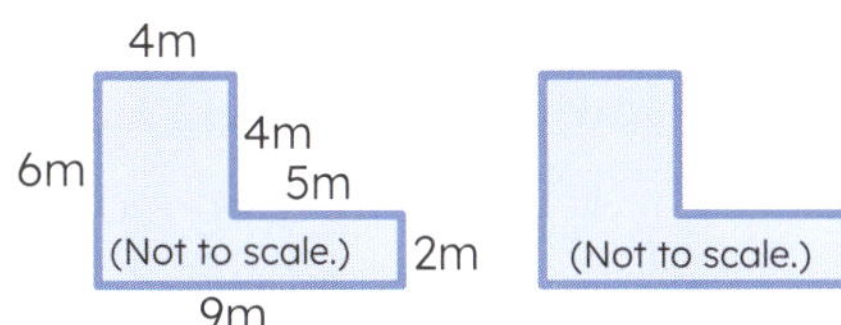

12. A coin-counting machine calculates by the mass of the coin. If a 10c coin weighs 5.65g, what is the value of 56.5g?

____

13. 100% = $\frac{100}{100}$ = _______

14. 100 000 – 31 000 = _______

15. 200 000 + _______ = 1 000 000

Thursday

Week 19

1. Continue the pattern.

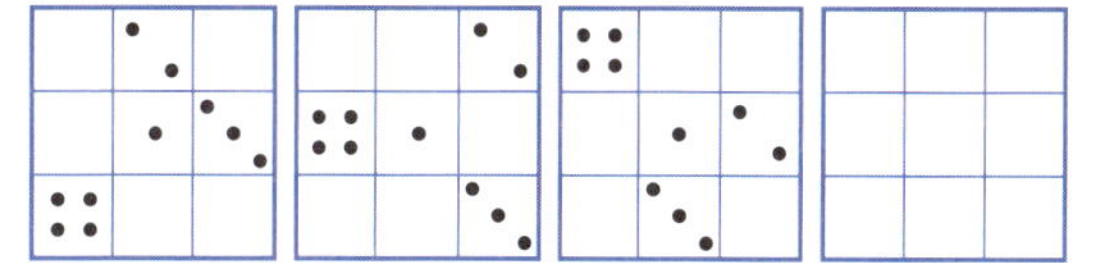

2. Simplify $\frac{10}{15}$. _______

3. What is the median of 5, 8, 10, and 12?

____

4. Name this shape. _______

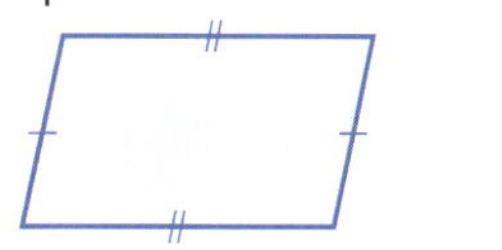

5. If $y \times 6 = 24$, then $y =$ _______.

6. 25% = (a) $\frac{1}{8}$ ☐ (b) $\frac{1}{4}$ ☐ (c) $\frac{1}{2}$ ☐ (d) 2.5 ☐

7. $\frac{3}{4}$ = 0._______ = _______%

8. Measure line $\overline{AB}$. _______mm

9. 45% of $1200 = _______

10. 32 000 + 158 000 = _______

11. Follow the instructions to correctly match $\frac{3}{4}$, 0.8, $1\frac{2}{3}$, and $\frac{19}{5}$ to each answer line.

(a) A fraction is left of the decimal.

(b) The mixed number is right of the improper fraction.

(c) The decimal is left of the improper fraction.

_______ _______ _______ _______

12. $50.00 – $31.65 = _______

13. What is the chance of randomly picking hearts or diamonds from a deck of 52 cards? Mark it on the scale.

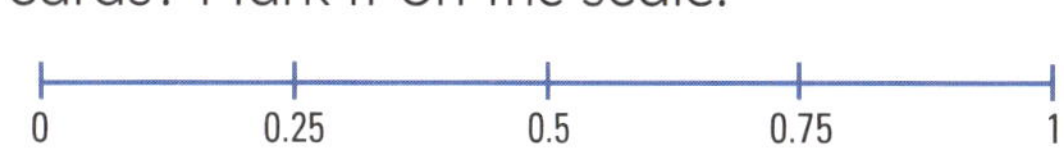

14. $8\frac{3}{5} + 1\frac{3}{5}$ = _______

15. 11^2 = _______

Week 19

Problem-solving

To train for a marathon, Johnny ran for 10 minutes on Monday, 15 minutes on Tuesday, 20 minutes on Wednesday, and increased the time by the same increment each day until they ran 55 minutes the following Wednesday.

What is the total time Johnny spent running across the 10 days, in minutes and hours/minutes?

Read the question again. **Think** about the information. Underline the important words.

Tick the strategy you will use to work out the answer:

- estimate and check ☐
- look for patterns ☐
- draw a diagram or picture ☐
- construct a table or graph ☐
- use materials ☐
- use a formula ☐
- something else. ☐

Solve it:

Reflect on the question and answer.

Check it. Circle another strategy on the list to work it out.

Show it:

Friday Review

1 What multiple of 3 is at A?

____

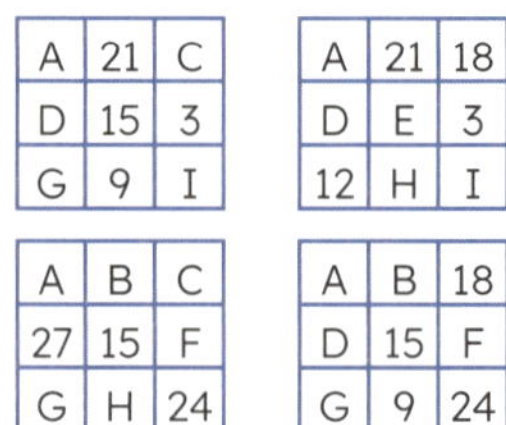

A	21	C
D	15	3
G	9	I

A	21	18
D	E	3
12	H	I

A	B	C
27	15	F
G	H	24

A	B	18
D	15	F
G	9	24

2 55% of $1400 =

____

3 1 000 000 − 15

= ________

4 42% = 0.________

$= \frac{\square}{100}$

5 A car is travelling constantly at 100km/h. If it makes no stops and travels 650km, how much time does the trip take?

____

6 Write one million, one thousand and ten as a numeral.

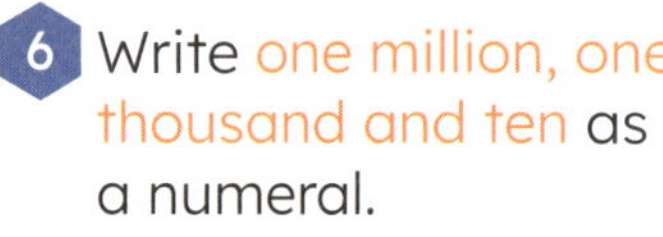

____

7 300 − 50 ÷ (50 ÷ 5)

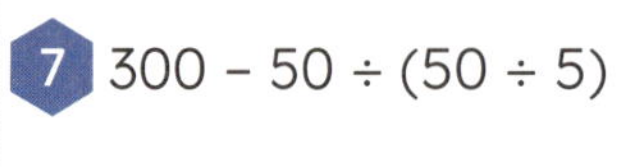

= ________

8 What is the median of 20, 30, 32, and 90?

____

9 What is the size of a°?

____

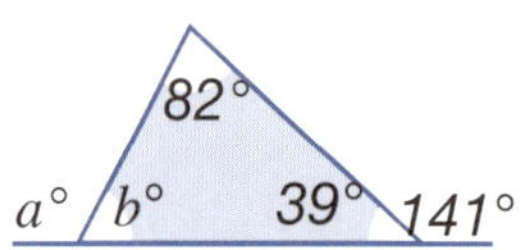

10 363 − 8 = ________

11 What is the probability of randomly selecting an even number card from a deck of 52 cards?

____

12 Draw the top view.

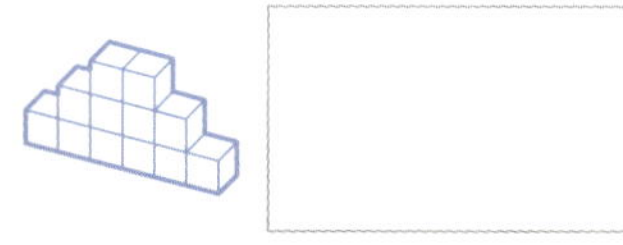

13 Is the bottom view identical to the top view?

yes ☐ no ☐

14 If a bus leaves its depot at 6:20 pm and stops six times with 12-minute intervals, what time does it reach the fourth stop?

____

15 $\frac{3}{4}$ = 0.________

16 100 − 0.005 = ________

17 6100mm = ________m

18 A rectangular-shaped warehouse has a perimeter of 90m. If the long side is 35m in length, the short side is

________.

N A M Sp St P

Monday

1.

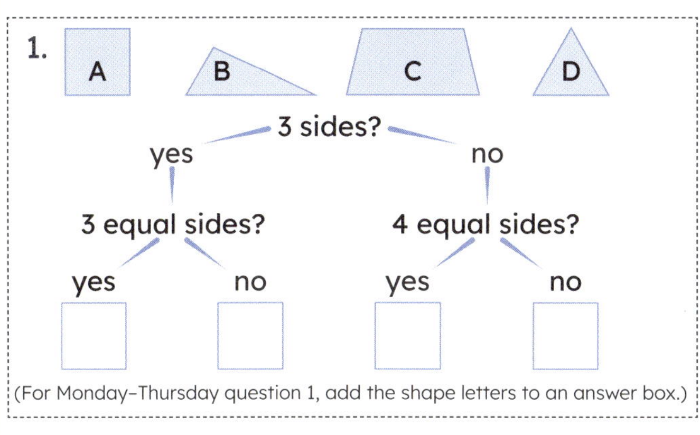

(For Monday–Thursday question 1, add the shape letters to an answer box.)

2. 24.17 = 20 + 4 + 0.________ + 0.________

3. $\frac{2}{5}$ is between: (a) 0.2 and 0.3. ☐
(b) 0.3 and 0.5. ☐
(c) 0.5 and 0.7. ☐

4. Name this shape. ________

5. 12.5% of \$200 = ________

6. Draw the right-side view.

7. The time is 4:30 am. What will it be in 10 hours? ________

8. Change $4\frac{3}{7}$ to an improper fraction. ________

9. Lucy's phone has half as many songs as Sam's phone. Which expression can be used to show the amount of songs on Lucy's phone? (s = Sam, l = Lucy)

(a) $s \times 2$ ☐ (b) $s \div 2$ ☐

(c) $s + (\frac{1}{2})^2$ ☐ (d) $l \div 2$ ☐

10. Add $1\frac{1}{2}$ hours to the time.

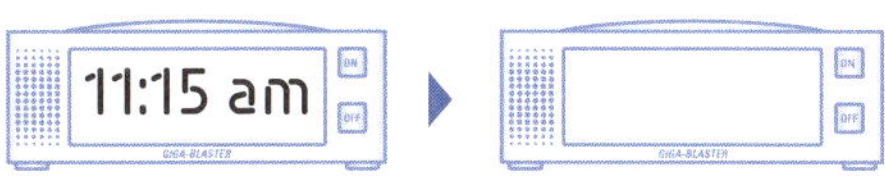

11. What is the radius of a biscuit tin with a 30cm diameter? ________

12. (a) 2 × 25 000 = ________

(b) 4 × 25 000 = ________

(c) 8 × 25 000 = ________

13. 230 + 480 = ________

14. You walked from A to B. How far did you walk? ________

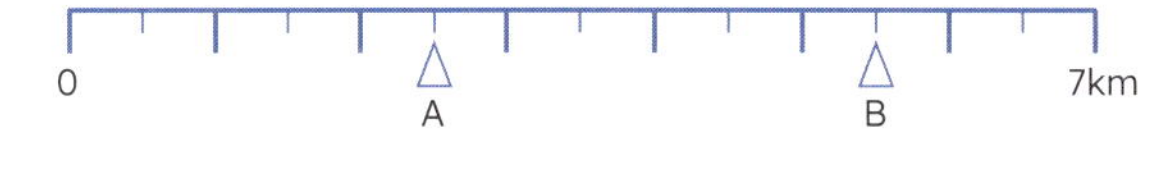

15. If 100 000 = 10a, then a = ________.

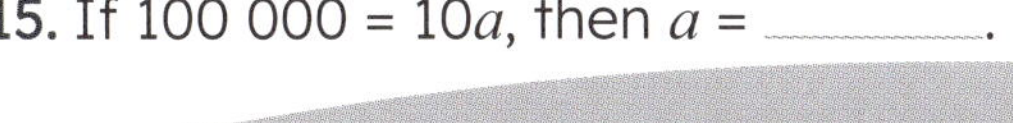

Tuesday

Week 20

1.

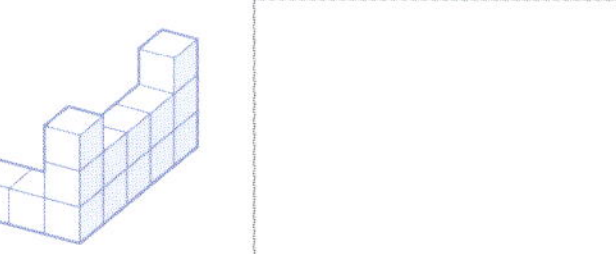

2. 285 000 + 70 000 = ________

3. 200 seconds = ________ min. ________ sec.

4. Add $9\frac{1}{2}$ hours to the time.

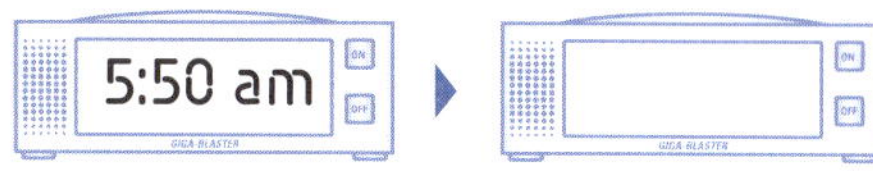

5. Double 269. ________

6. The LCD for $\frac{3}{8}$ and $\frac{2}{3}$ is ________.

7. Alicia and Beau walked from A to B. How far did they walk? ________

0 A B 12km

8. What is the product of 6 and 9? ________

9. The clock's hour hand fell off 10 minutes before 6 o'clock. Draw the hour hand in the correct position.

10. If 40 + y = 130, then y = ________.

11. If there are 10 000 Collinghood fans and 1000 Bullfrogs football fans, what is the simplified ratio of fans? ________

12. Round π to the nearest whole. ________

13. Write 8.195 million as a numeral. ________

14. Mr Wallet counted and weighed a pile of coins. Complete the table.

Single Coin	Mass (grams)
5c	2.83
10c	5.65
20c	11.3
50c	15.5
\$1	9
\$2	6.6

Coin	Mass (grams)	Value
5c	28.3	0.50
10c	5.65	0.10
20c	33.9	(a)
50c	155	(b)
\$1	(c)	9.00
\$2	(d)	6.00
Total value (e) \$		

15. 0.35 = ________%

Week 20

Wednesday

1. A B C D

Is it a hexagon?

yes — regular shape? yes / no

no — irregular shape? yes / no

2. What is the product of 8 and 4? ______

3. If $a - 30 = 50$, then $a =$ ______.

4. $\frac{31}{9} =$ ______

5. Which two prime numbers are the sum of 32?

______ , ______

6. 29 – 7 = ______

7. $a^\circ =$ ______

8. 17 + 5 × 6 = ______

9. 10kL = ______ L

10. Draw the net of a triangular prism.

11. \$5.00 – \$1.20 = ______

12. 8 ⟌408 = ______

13. 65% of \$300 = ______

14. Shayan (s) has twice as many marbles as Matthew (m). Which expression shows the number of marbles Shayan has?

(a) $s \div 2$ ☐ (b) $m \times 2$ ☐

(c) $s \times 2$ ☐ (d) $s + 2$ ☐

15. Order the boxes of cereal from lightest to heaviest.

______ ______ ______

A 

B 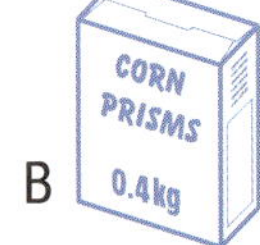

C 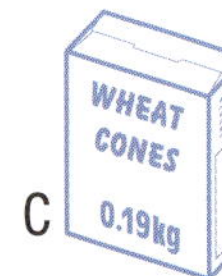

Thursday

1.

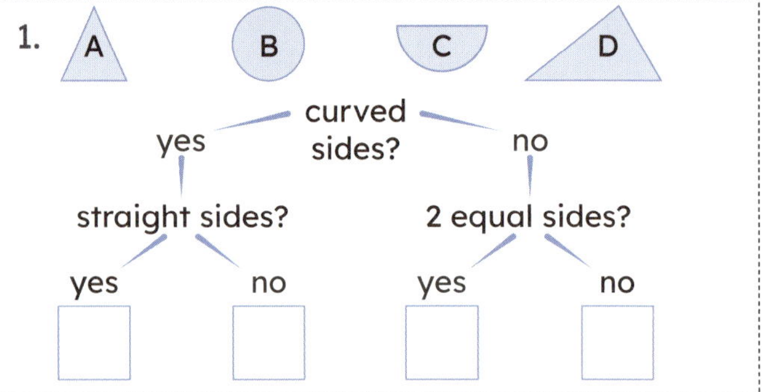

2. What is the perimeter of a 7cm regular pentagon? ______

3. Which is the acute angle? ______

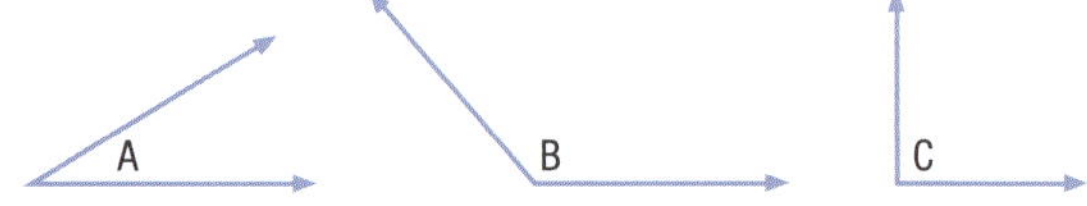

4. Flip a coin. The two outcomes are ______ or ______.

5. $\frac{3}{4} = 0.75$ true ☐ false ☐

6. A pop star, Tiffy Pearl, has a pathway made of rose petals. It takes 2.5kg of petals to cover 1m^2. The path to the stage is 2m wide and 8m in length. How many kilograms of rose petals are needed?

______

7. 94 – 9 = ______ , 9.4 – 0.9 = ______

8. 40 – 0.08 = ______

9. Name this type of triangle. ______

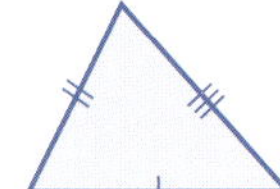

10. Is 1735 divisible by 3? yes ☐ no ☐

11. If $2 \times a = 400$, then $a =$ ______.

12. 1 150 000 + 250 000 = ______

13. What is the difference between 101 and 10?

______

14. \$50.00 – \$29.35 = ______

15. Which set of 4 are prime numbers?

(a) 2, 5, 9, 11 ☐ (b) 3, 6, 9, 11 ☐

(c) 2, 4, 6, 10 ☐ (d) 2, 5, 7, 19 ☐

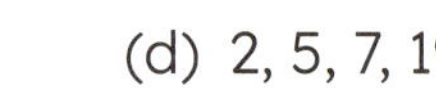

Problem-solving

Consider the side and angle properties of the following three- and four-sided shapes: equilateral triangle, isosceles triangle, scalene triangle, square, rectangle, rhombus, parallelogram, kite, and trapezium.

Create a tree diagram, like the ones used for this week's focus questions, to sort these nine shapes.

Read the question again. **Think** about the information. Underline the important words.

Tick the strategy you will use to work out the answer:

- estimate and check ☐
- look for patterns ☐
- draw a diagram or picture ☐
- construct a table or graph ☐
- use materials ☐
- use a formula ☐
- something else. ☐

Solve it:

Reflect on the question and answer.

Check it. Circle another strategy on the list to work it out.

Show it:

Friday Review

1 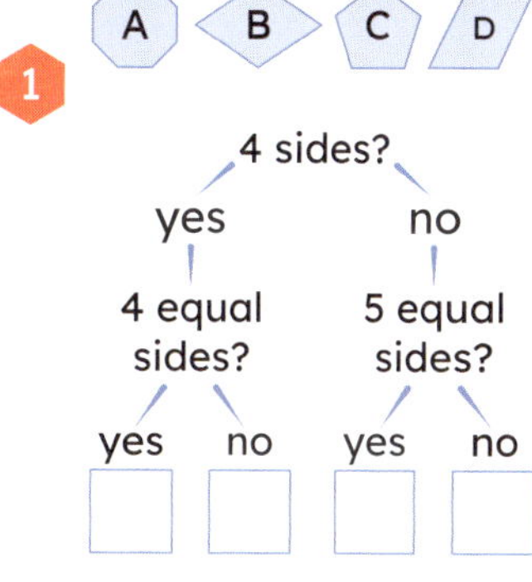

(Add the shape letters to an answer box.)

2 0.45 = ________ %

3 A bush tucker store has twice as many desert limes (d) than bunya nuts (b). Which expression shows the number of bunya nuts?

(a) $d \div 2$ ☐

(b) $d \times 2$ ☐

(c) $b \div 2$ ☐

(d) $b \times \frac{1}{2}$ ☐

4 The product of 6 and 8 is ________.

5 Double 780.

____

6 Write 6.237 million as a numeral.

____

7 $9\frac{3}{4} + \frac{1}{2} =$ ________

8 $\frac{3}{5}$ is between:

(a) 0.3 and 0.5. ☐

(b) 0.5 and 0.7. ☐

(c) 0.7 and 0.9. ☐

9 85% of 800 = ________

10 Which is not a net of a triangular prism?

____

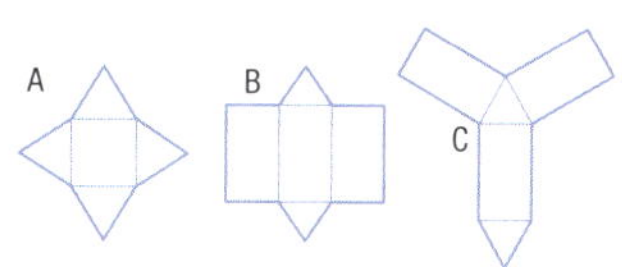

11 9.97, 9.98, 9.99,

____

12 Is 1935 divisible by 9?

____

13 If $2 \times a = 1000$, then

$a =$ ________.

14 17 500mm

= ________ m

15 If there are 8000 Newcastle Magpies fans and 4000 Adelaide Devils fans, what is the simplified ratio?

____

16 What is the perimeter of a regular hexagon with 9cm sides?

____

17 Which box is the lightest?

____

18 What is the probability of randomly selecting a green jelly bean from a jar if there are 50 green, 100 red, and 100 white jelly beans?

____

Week 21

Monday

1. Continue the powers of two sequence:

   1, 2, 4, 8, ________, ________, ________

2. This is a net of a ________.

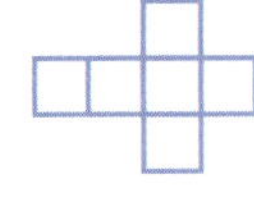

3. Measure $\overline{CD}$. ________mm

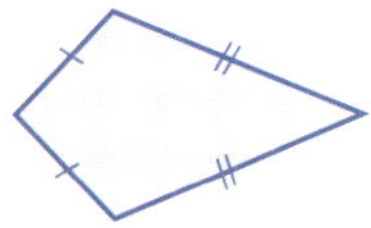

4. 70, 140, 210, ________, 350, ________

5. Double 950. ________

6. There are 150 more beans (b) in red jars (r) than in green jars (g). Which expression shows the number of beans in a red jar?

   (a) $r + 150$ ☐ (b) $r - 150$ ☐ (c) $g + 150$ ☐

7. $\frac{2}{3} + \frac{2}{3} =$ ________

8. Name this shape. ________

9. Add brackets to make this number sentence true.

   7 ÷ 1 – 5 + 2 = 0

10. 40 × 50 = 20 × ________

11. Round 6.78 to the nearest tenth. ________

12. Which is the obtuse angle? ________

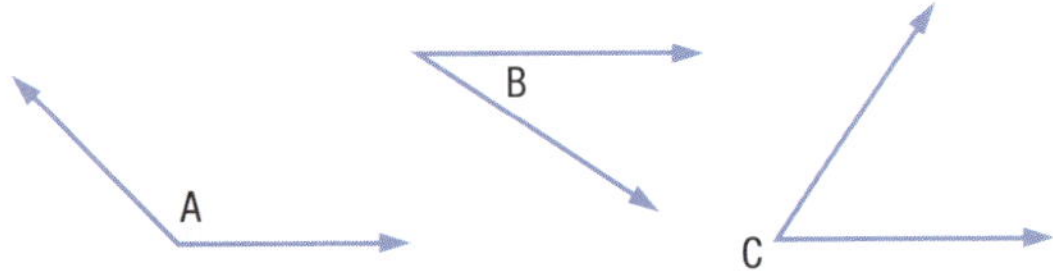

13. Simplify 8:12. ________

14. What are the prime factors of 36?

    ____

15. These lines are:

   parallel. ☐ parallelograms. ☐

   pararails. ☐ two wickets. ☐

Tuesday

1. Continue the powers of three sequence:

   1, 3, 9, 27, ________, ________, ________

2. Which decimal is closest to 5.72?

   (a) 5.718 ☐ (b) 5.725 ☐

   (c) 5.71 ☐ (d) 5.7 ☐

3. If $40 \times y = 200$, then $y =$ ________.

4. Change $4\frac{3}{7}$ to an improper fraction. ________

5. How many B boxes will fit into box A?

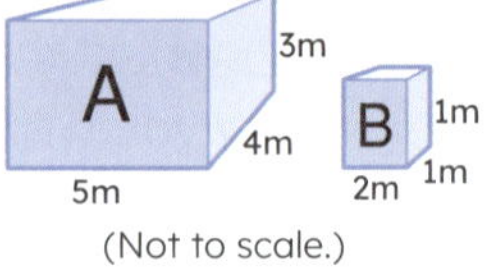

(Not to scale.)

   ____

6. Write 4:20 pm in 24-hour time. ________

7. A water tank has a capacity of 12kL but is only one-third full. How many kilolitres are in the tank?

   ____

8. $\frac{4}{5} = 0.$________ $= \frac{\square}{100}$

9. How many $2.00 coins make up $500.00?

   ____

10. Which line is perpendicular to x? (w, y, or z.)

    ____

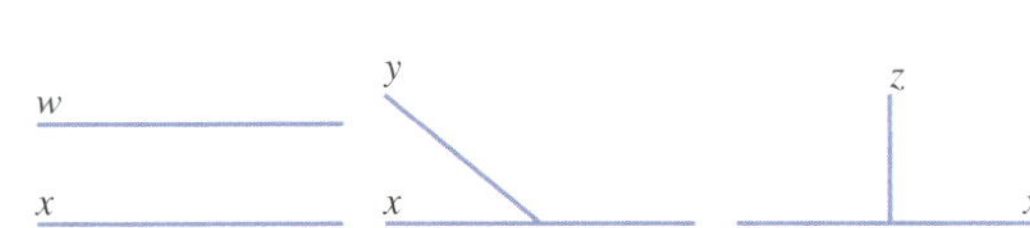

11. What is the probability of randomly picking an odd number from a pack of 52 playing cards?

    ____

12. A class of Year 7s has 8 students with the following shoe sizes: 6, 8, 5, 4, 9, 12, 5, 7.

    What is the mean shoe size?

    ____

13. 40% of $90.00 is ________.

14. Which letter is opposite E?

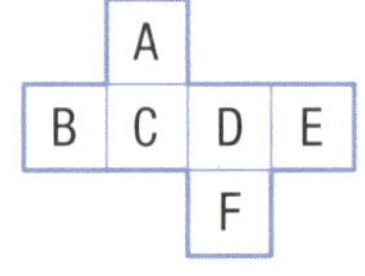

    ____

15. 8500 – 700 = ________

Wednesday

1. Continue the powers of four sequence:

   1, 4, 16, ________, ________, ________

2. 20 + 7 × 2 = ________

3. 5 – 0.005 = ________, 0.5 – 0.005 = ________

4. What is the sum of the internal angles of a triangle?

   ____

5. What speed is a car travelling at if it covers 5km every 5 minutes?

   ____

6. Halve 900. ________

7. A water tank has a capacity of 15kL. If it is two-thirds full, how much water is needed to fill the tank?

   ____

8. What is the mode of the following numbers?

   6, 7, 8, 7, 5, 7, 9, 7

   ____

9. The value of 4 in 41 789 is ________.

10. Draw a reflection.

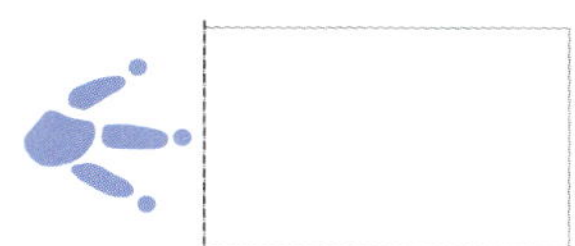

11. $10\overline{)1000}$ = ________

12. 1.007kg = ________g

13. Which day's temperature was an anomaly (outlier) for the week?

    ____

Brisbane's seven day temperature (°C)						
Mon	Tues	Wed	Thurs	Fri	Sat	Sun
21	20	12	19	22	23	20

____

14. $\frac{2}{7}$ of 2135 = ________

15. 30% of $1500 = ________

Thursday

1. Continue the powers of four sequence:

   1, 5, 25, 125, ________, ________, ________

2. What do the internal angles of any quadrilateral add up to?

   ____

3. Name this shape. ________

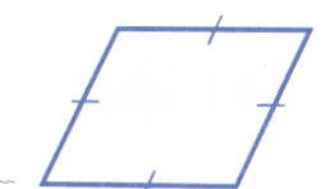

4. A shipping container carries boxes of plasma TVs. How many TV boxes will fit into this sea container?

   ____

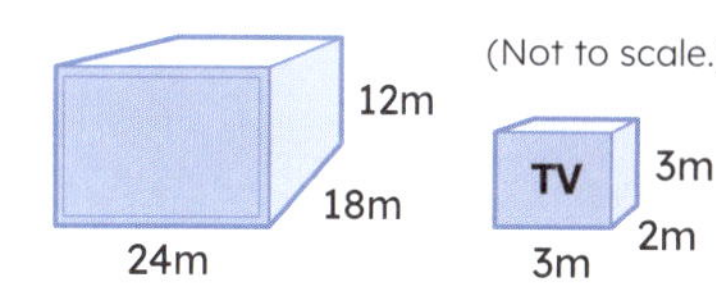

5. Round 8.447 to the nearest hundredth.

   ____

6. Is 179 divisible by 9? ________

7.

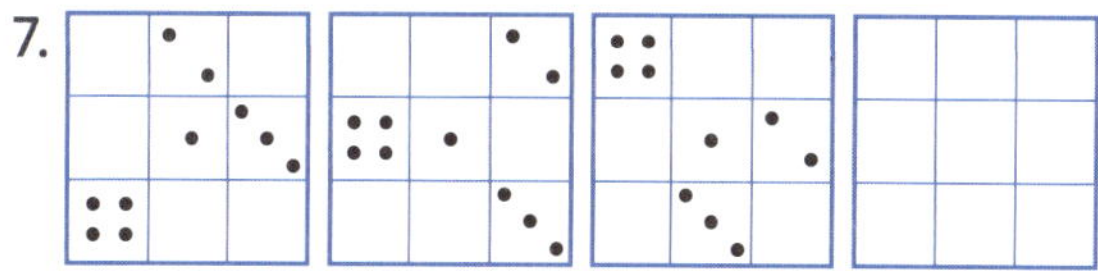

8. 4.2m = ________cm

9. Write in ascending order: 0.09, $\frac{5}{8}$, 0.5, 1.2.

   ________, ________, ________, ________

10. Does this First Nations' star symbol have rotational symmetry?

yes ☐ no ☐

11. If 7 × 30 = y + 10, then y = ________.

12. Area of triangle = ________cm^2

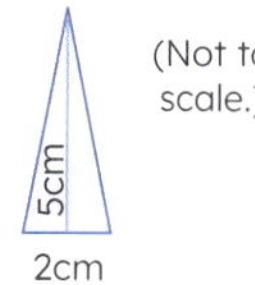

(Not to scale.)

13. Lyla's race car time was 10.24 seconds. Blake clocked 10.78 seconds. Who was faster and by how much?

    ____

14. 15% of $2100 = ________

15. $\frac{4}{5} + \frac{4}{5}$ = ________

Problem-solving

When a number (base) is raised to powers increasing by 1 each time, it creates a sequence. For example, using a base of 3 produces the sequence 1, 3, 9, 27, 81, 243, 729, etc.

Which of the numbers 16, 64, 256, or 512 appear in the sequences produced when 2, 4, 6, or 8 are raised to increasing powers?

Read the question again. **Think** about the information. Underline the important words.

Tick the strategy you will use to work out the answer:

- estimate and check ☐
- look for patterns ☐
- draw a diagram or picture ☐
- construct a table or graph ☐
- use materials ☐
- use a formula ☐
- something else. ☐

Solve it:

Reflect on the question and answer.

Check it. Circle another strategy on the list to work it out.

Show it:

Friday Review

1. $\frac{2}{7}$ of 2842 = ________
2. 20% of $60.00 = ________
3. Double 850. ________
4. A water tank has a capacity of 18kL. Two-thirds of the tank is empty. How many kL of water is currently in the tank? ________
5. Area of triangle = ________cm^2

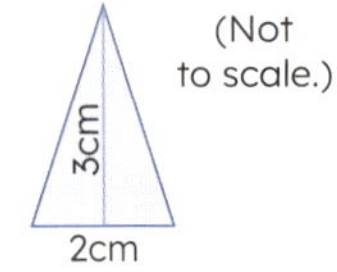

(Not to scale.)

6. How many degrees make up a square? ________
7. If 70 × a = 210, then a = ________.
8. Measure line $\overline{XY}$. ________mm

X —————— Y

9. Round 524 690 to the nearest hundred thousand. ________
10. Continue the powers of five sequence: 1, 5, ________, 125, ________, ________
11. 1kL = ________L
12. What is the mode of 6, 8, 16, 6, and 10? ________
13. A class of Year 7s has 10 students with the following shoe sizes: 12, 8, 5, 4, 9, 12, 5, 7, 12, 6.

    What is the mean shoe size? ________
14. $\frac{3}{5}$ = 0.________ = $\frac{\square}{100}$
15. What is the probability of randomly picking a king from a pack of 52 playing cards? ________
16. Does this shape have rotational symmetry?

    yes ☐ no ☐

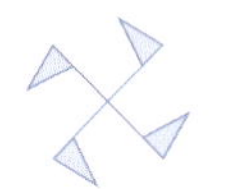

17. 3.375km = ________m
18. This is a net of a ________.

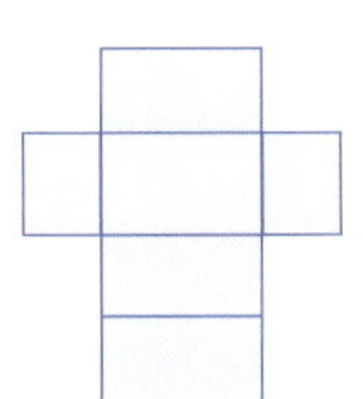

St P

Monday

1. Which temperature is the anomaly?

M	Tu	W	Th	F
20°C	21°C	22°C	21°C	34°C

2. What do the internal angles of a triangle sum to?

3. What is the sum of 90 and 75?

4. A jar holds 20 red and 30 blue lollies. What is the probability of randomly selecting a blue lolly?

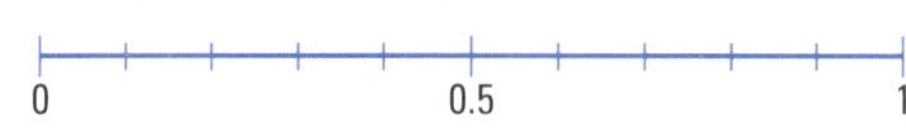

5. 3^3 = ______

6. 0.97, 0.98, 0.99, ______

7. 42 ÷ ______ = 7

8. What is the mean: 60, 80, 50, 90, 70?

9. Simplify the ratio 8:24. ______

10. The value of 4 in 2.004 is ______.

11. Name this 3D object.

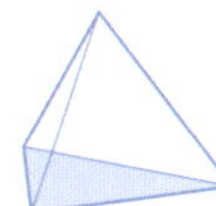

12. Mila exchanged A$25 for EUR€29. How many euros will Mila get with A$100?

13. Draw a pair of parallel lines.

14. Which prime numbers are located between 40 and 50?

15. Over two hours, Antonio packed 300 bananas into bags. Each bag had five bananas. How many bags of bananas are there? Which expression matches?

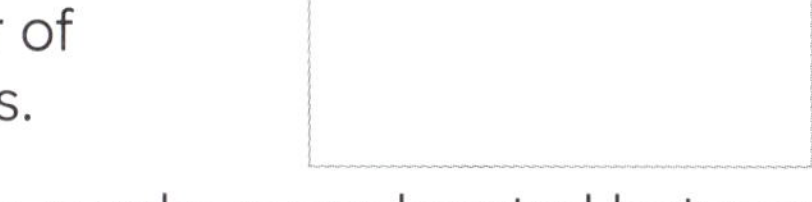

(a) 300 ÷ 5 ☐ (b) 300 – 5 ☐

(c) 5 ÷ 300 ☐ (d) 300 ÷ 2 ☐

Tuesday

Week 22

1. Which amount of rainfall is the anomaly?

M	Tu	W	Th	F
4mm	21mm	23mm	19mm	22mm

2. 109 014 – 20 = ______

3. Write in descending order: 0.9, 2.4, $\frac{1}{5}$, $1\frac{1}{5}$, $\frac{3}{10}$.

______, ______, ______, ______, ______

4. If a car travels at 100km/h, how far will it travel in 15 minutes?

5. 2^6 = ______

6. $\frac{7}{10} - \frac{2}{10}$ = ______

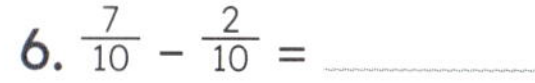

7. Measure $\overline{ABC}$. ______mm

8. Rotate the shape on its • by a quarter turn clockwise.

9. If chocolate costs $4.95 per 100g, what is the cost of one kilogram?

10. $\frac{3}{7}$ × 3521 = ______

11. Write the numeral 10 before ten million, one thousand and one.

12. After making a cube from this net, which face is opposite B?

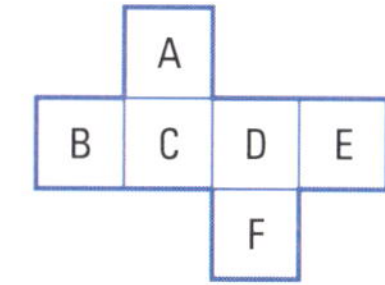

13. $a°$ = ______

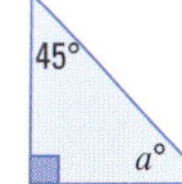

14. 8.2 ÷ 10 = ______

15. The formula for the area of a rectangle is

______ × ______.

Week 22

Wednesday

M	Tu	W	Th	F
5°C	4°C	14°C	3°C	4°C

1. Which temperature is the anomaly?

______

2. What is the range and mean of the temperatures above?

Range = ______ Mean = ______

3. Write 6:45 pm as 24-hour time. ______

4. If it is 8:30 pm Tuesday, what will be the day and time in 26 hours?

______

5. Write in ascending order: $\frac{1}{10}$, $\frac{3}{4}$, $\frac{1}{2}$, $\frac{2}{100}$.

______, ______, ______, ______

6. 7 ⟌4907 = ______

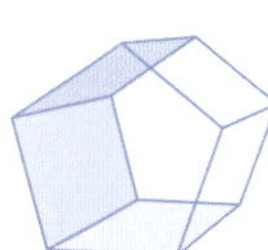

7. Name this 3D object.

______

8. 31, 34, 37, ______, 43

9. If a deli charges $2.45 per 100g for cheese, what is the cost per kg?

______

10. What is the value of:

(a) 45g of $1?

______

(b) 113g of 20c?

______

Single coin	Mass (grams)
5c	2.83
10c	5.65
20c	11.3
50c	15.5
$1	9
$2	6.6

11. Write the prime factors of 15. ______

12. 7 × 6 = ______

13. If 5000 – a = 400 × 5, then a = ______.

14. There are 360 adults and children at St James' Park. The ratio of adults to children is 1:3. How many adults are there?

(a) 90 ☐ (b) 60 ☐

(c) 360 ☐ (d) 120 ☐

15. How many $2.00 coins make up $1002?

______

Thursday

M	Tu	W	Th	F
12°C	14°C	12°C	1°C	11°C

1. Which temperature is the anomaly?

______

2. What is the range and mean of the rainfall above?

Range = ______ Mean = ______

3. A trapezium has 4 internal angles that add up to ______°.

4. 6.95 ÷ 10 = ______

5. 8.85, 8.9, 8.95, ______

6. 8 ⟌6408 = ______

7. Look at Wednesday's table. What is the value of:

(a) 283g of 5c?

______

(b) 139.5g of 50c?

______

8. $\frac{8}{20} + \frac{14}{20}$ = ______

9. 9 – 0.007 = ______

10. Draw each shape's diagonals and write its amount of lines. (Do you see a pattern?)

______ ______ ______

 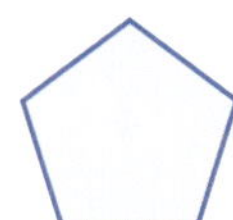 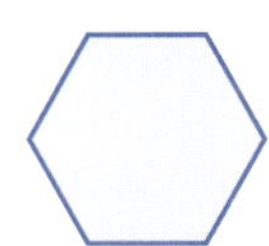

11. The formula for the area of a circle is

______.

12. Mia paid $200 for new jeans. Pia paid 10% more for the same jeans at another store. What did Pia pay?

______

y° 50° 50°

13. $y°$ = ______

14. 3.965kL = ______L

15. 2.076 ÷ ______ = 0.2076

Problem-solving

Three students spent five days collecting witchetty grubs from habitats near their school.

	Number found among the witchetty plants	Number found among the acacia shrubs	Number found among the black wattle trees
Mon	56	24	39
Tues	59	20	27
Wed	53	25	41
Thurs	54	36	38
Fri	58	23	40

Can you identify the two anomalies (outliers) in the data? (Hint: Display the table's data in a different format, so the two anomalies can be clearly seen.)

Read the question again. **Think** about the information. Underline the important words.

Tick the strategy you will use to work out the answer:

- estimate and check ☐
- look for patterns ☐
- draw a diagram or picture ☐
- construct a table or graph ☐
- use materials ☐
- use a formula ☐
- something else. ☐

Solve it:

Reflect on the question and answer.

Check it. Circle another strategy on the list to work it out.

Show it:

Friday Review

Week 22

M	Tu	W	Th	F
7.5°C	8°C	15°C	7°C	7.5°C

1. Which temperature is the anomaly? ______

2. What is the range and mean of the temperatures above?

   Range = ______

   Mean = ______

3. $a° =$ ______

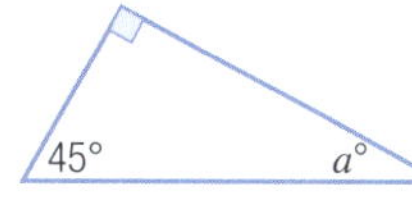

4. Name this shape. ______

5. If $6000 - a = 300 \times 10$, then $a =$ ______.

6. 7.85, 7.90, 7.95, ______

7. $\frac{6}{20} + 1\frac{19}{20} =$ ______

8. $2^5 =$ ______

9. Double 697. ______

10. Look at Wednesday's table. What is the value of 226g of 20c? ______

11. If the radius of a circle is 60cm, what is its diameter? ______

12. $11^2 =$ (a) 121×2 ☐ (b) 11×11 ☐ (c) 22 ☐

13. $\frac{9}{10} - \frac{5}{10} =$ ______

14. Write the prime factors of 18. ______

15. If a deli charges $2.10 per 100g for smelly cheese, what is the cost per kg? ______

16. Which is heaviest?

    (a) 0.07kg ☐

    (b) 18kg ☐

    (c) 10 000g ☐

17. If a blue car is travelling at 80km/h, how far will it travel in 1 hour? ______

18. What is the probability of selecting a green jelly sweet from a jar which has 20 yellow, 20 green, and 60 red jellies? No looking! ______

Week 23

Monday

1. Kylie ate toast for breakfast for four out seven days last week. How many times would you expect Kylie to eat toast in six weeks?

20 ☐ 24 ☐ 28 ☐

2. $y°$ = ______

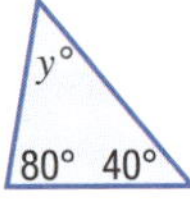

3. If $6\frac{1}{2} + e = 6\frac{3}{4}$, then e = ______.

4. Which shape has the same fraction shaded as A? ______

A  B 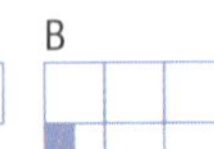 C 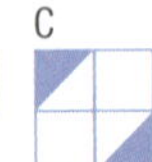 D 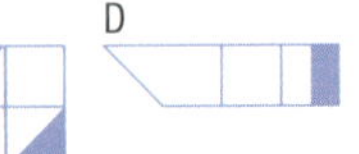 E 

5. Round 1.459 to the nearest tenth.

______

6. If it is 2:30 am Saturday, what will be the day and time in 30 hours?

______

7. 44, 88, 176, ______

8. 3050m = ______km

9. 2 × 3 × 3 × 3 are the prime factors for:

______.

10. $\frac{4}{7}$ of 2828 = ______

11. (40 ÷ 5) × (5 × 10) = ______

12. Double 8.9. ______

13. A farm's square paddock has four fence posts on each side. How many posts are there altogether?

______

14. If you had $10.80, of which $5.80 was in 20c coins and the balance in 50c coins, how many coins are there altogether?

______

15. Which line is perpendicular to E? ______

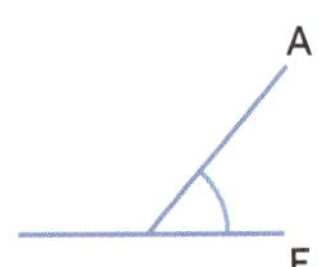

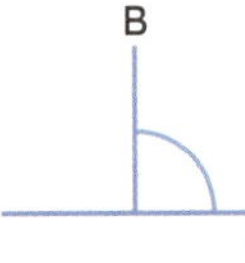

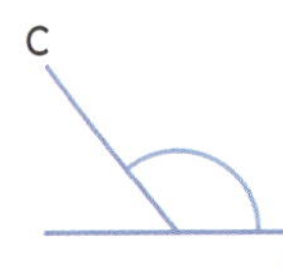

Tuesday

1. 25 animals came to the waterhole on Monday. How many animals would you expect to come for a drink during the entire week?

175 ☐ 225 ☐ 250 ☐

2. What is the difference between 80 and 35? ______

3. What is the starting number (x)? ______

x × 3 ➡ ______ ÷ 2 ➡ ______ + 20 ➡

______ – 50 = 30

4. If a car travels 5km in 3 minutes, how far could it travel in one hour? ______

5. Write 12:30 am in 24-hour time. ______

6. $\frac{3}{5}$ = 0.______

7. Which 3D object has half as many faces as edges? ______

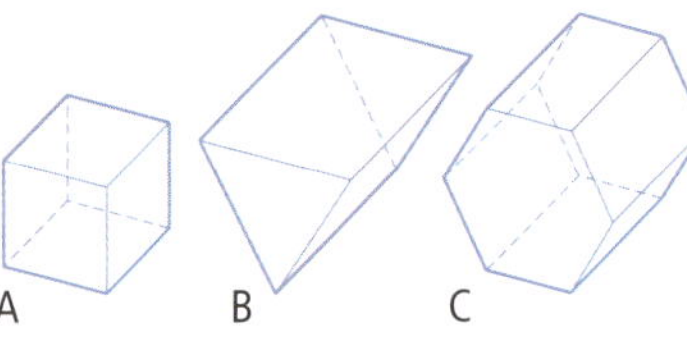

8. What is the chance of you randomly selecting a king from a deck of 52 playing cards?

______

9. 9.7 ☐ 10 = 0.97

10. If your mum paid $12.00 per kg for rice, what is the cost of $3\frac{1}{2}$kg? ______

11. Halve 12.7. ______

12. If it is 4:50 pm Sunday, what will be the day and time $20\frac{1}{2}$ hours later?

______

13. Colin (c), a car collector, has three times as many cars as the local museum (m). Which expression shows the number of cars Colin has?

(a) $m + 3$ ☐ (b) $m \times 3$ ☐ (c) $c \div 3$ ☐

14. 6207g = ______kg

15. Write in descending order: 2.02, 2.0, 2.2, 2.002.

______ ______ ______ ______

Wednesday

1. A bakery sold 60 types of bread and 32 types of cake on Monday. How many items would you expect the bakery to sell in 10 days?

   860 ☐ 890 ☐ 920 ☐

2. 200, 20, 2, ________
3. (700 ÷ 7) ÷ 2 = ________
4. $(9 \times 10^5) + (9 \times 10^3) =$ ________
5. 95% of 1700 = ________
6. 75% of $300.00 is ________.
7. $y°$ = ________

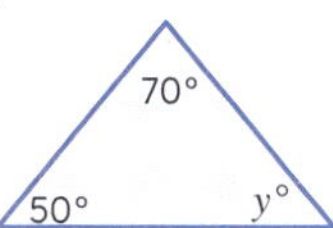

8. 100 000 – 40 000 = ________
9. If $2\frac{3}{5} + e = 3\frac{1}{5}$, then e = ________.
10. 10 000 = ________ × ________ × ________ × ________ $= 10^4$
11. 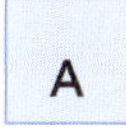  

3 sides?
yes — 3 equal sides? (yes / no)
no — 4 equal sides? (yes / no)

 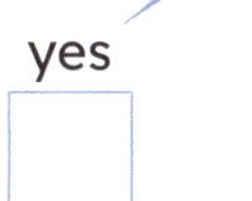 

(Add the shape letters to an answer box.)

12. Write ten million, one hundred as a numeral.

    ____

13. Which pair is parallel? ________

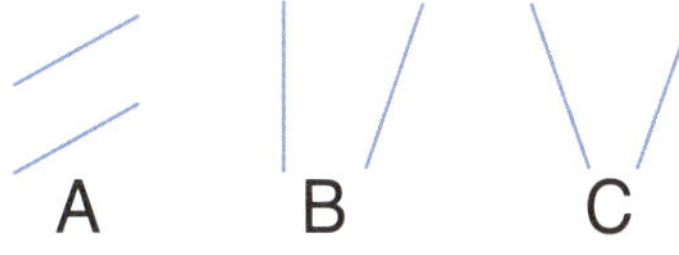

14. What is the perimeter of a regular hexagon with 70mm sides?

    ____

15. Round 11 502 to the nearest thousand.

    ____

Thursday

1. Brendan received $5.50 in pocket money each week. How much money would you expect Brendan to receive in eight weeks?

   $38.59 ☐ $41.25 ☐ $44.00 ☐

2. $(7 \times 10^4) + (2 \times 10^3) + (4 \times 10^2) =$

   ____

3. The angle size of a straight line is ________.
4. 18 + 15 = 3 × ________
5. If you throw a die, the chance of it landing on a six is 1 in 6. If you throw two dice, what is the chance of throwing two sixes?

   ________ in ________

6. $1\frac{3}{5} + \frac{4}{5} =$ ________
7. The angle at x = ________, and y = ________.

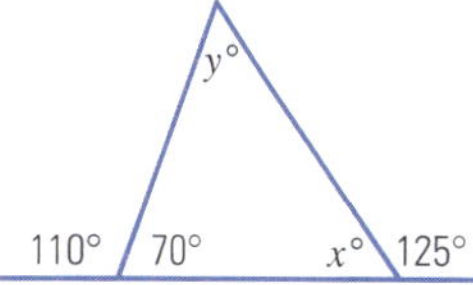

8. $\frac{3}{5}$ of 2025 = ________
9. $3\frac{1}{5} - \frac{4}{5} =$ ________
10. 7093mm = ________m
11. 9 × 2 – 1 = ________
12. Draw a shape double the size of this shape.

13. A circular driveway has a 5m radius. What is the diameter?

    ____

14. The value of 7 in 2.007 is ________.
15. If 22 ÷ n = 5.5, then n = ________.

Week 23

Problem-solving

Ellin, Kirra, and Arika were competing in an apwerte challenge. The aim is to roll a stone through two markers. After 20 turns, Ellin's score was 5, Kirra's 7, and Arika's 9. After 50 turns, Ellin's score was 12, Kirra's 18, and Arika's 24.

What would you expect their scores to be after 100 and 200 turns each? Why?

Read the question again. **Think** about the information. Underline the important words.

Tick the strategy you will use to work out the answer:

- estimate and check ☐
- look for patterns ☐
- draw a diagram or picture ☐
- construct a table or graph ☐
- use materials ☐
- use a formula ☐
- something else. ☐

Solve it:

Reflect on the question and answer.

Check it. Circle another strategy on the list to work it out.

Show it:

Friday Review

1. Lucy wore red socks for five out of seven days last week. How many times would you expect Lucy to wear green socks in six weeks?

   20 ☐ 25 ☐ 30 ☐

2. 17 + 19 = 4 × ______

3. $5\frac{1}{5} - \frac{3}{5} =$ ______

4. $6\frac{4}{7} + 3\frac{5}{7} =$ ______

5. Pine Farm (a) has twice as many cows as Oak Farm (b). Which expression shows the number of cows at Oak Farm?

   (a) $a \div 2$ ☐

   (b) $b \times 2$ ☐

   (c) $a \times 2$ ☐

6. Double the size of this shape.

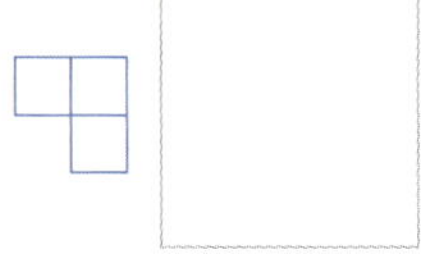

7. 3.8 ☐ 10 = 0.38

8. If you paid $10 per kilogram for garlic, what is the cost of $6\frac{1}{4}$kg?

   ______

9. 496 × 10 = ______

10. $y°$ = ______

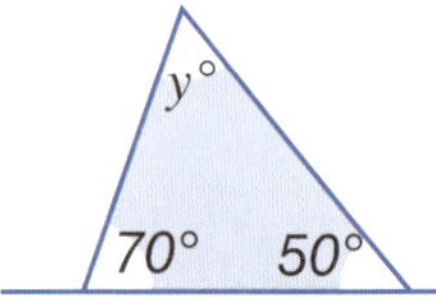

11. Write in descending order: $\frac{1}{5}$, $\frac{1}{3}$, $\frac{3}{10}$, $\frac{1}{100}$.

    ______, ______, ______, ______

12. Sam saw six flies, two ladybugs, and four bees in the garden on Monday. How many insects would you expect Sam to see in two weeks?

    168 ☐ 188 ☐ 208 ☐

13. Round 4.477 to the nearest hundredth.

    ______

14. 7 + 4 × 4 – 2 = ______

15. (400 ÷ 40) ÷ 5

    = ______

16. Name this 3D object.

    ______

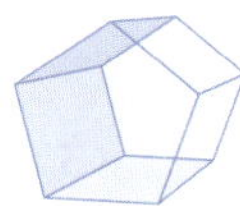

17. Write 2:05 am in 24-hour time.

    ______

18. If a car travels $4\frac{1}{2}$km in 3 minutes, how far can it travel in 1 hour?

    ______

N A M Sp St P

Monday

1. Parallelogram area = base × height =

______ cm × ______ cm = ______ cm^2

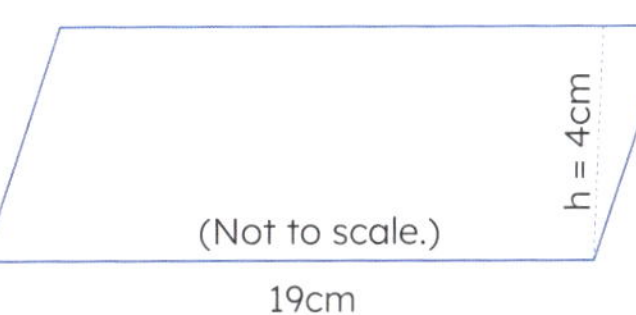

2. Round 0.297 to the nearest tenth. ______

3. Convert $9\frac{2}{3}$ to an improper fraction. ______

4. 2 – 1.75 = ______, 0.2 – 0.175 = ______

5. 0.50, 0.75, ______, 1.25

6. Vincenzo, a grocer, packed 5 bananas in each bag. There were 29 bags. Which expression matches how many bananas were packed altogether?

(a) 5 × 29 ☐ (b) 5 ÷ 29 ☐

(c) 29 – 5 ☐ (d) 5 + 29 ☐

7. If it is 6:35 pm Friday, what will be the day and time 46 hours later?

______

8. Double the size of this shape.

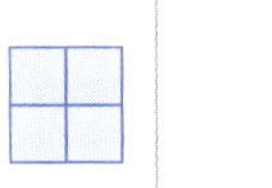

9. $6\frac{3}{5} + 2\frac{4}{5}$ = ______

10. 8979L = ______ kL

11. $y°$ = ______

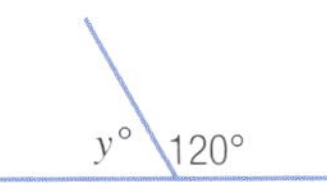

12. If $\frac{2}{3}$ of a class of 30 are wearing a uniform, how many are not wearing it? ______

13. 4 × 19 × 50 = ______

14. Continue the pattern.

27		9
	81	
	54	

	18	
36	81	
		63

27		
	81	72
45		

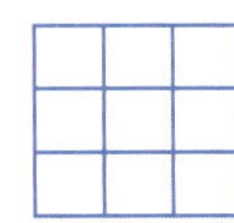

15. A game costs $60.00 and is discounted by 20%, what is its new price?

______

Tuesday

Week 24

1. Parallelogram area = $b \times h$ =

______ cm × ______ cm = ______ cm^2

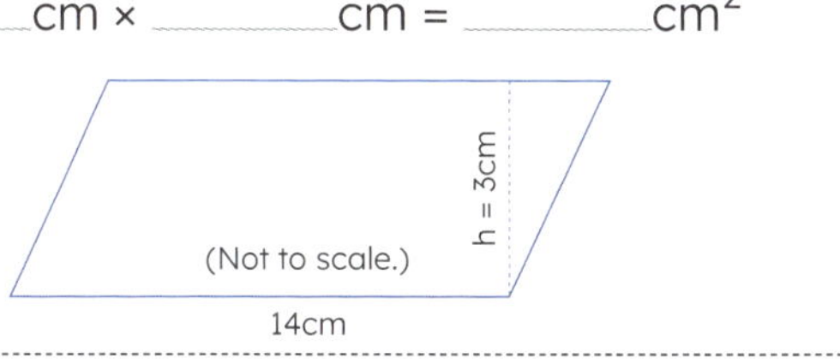

2. Daryl wants to grow grass and flowers in a small garden.

Each square = $2m^2$. 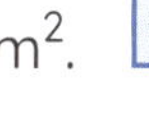 grass  roses 

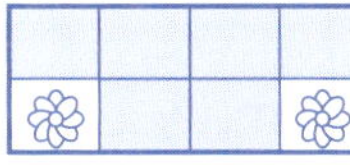

How many square metres of grass are needed? ______

3. If 5 × e = 400, then e = ______.

4. Simplify $\frac{30}{35}$. ______

5. Draw the top view.

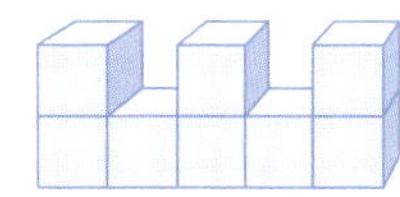

6. If you ride your new blue bike for half an hour and cover $13\frac{1}{2}$km, how fast are you travelling?

______ km/h

7. Name this 3D object.

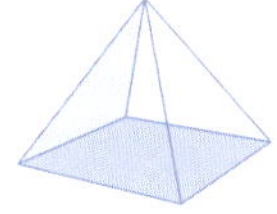

______

8. (a) Measure $\overline{AB}$. ______ mm

(b) Measure $\overline{AC}$. ______ mm

A B C

9. $\frac{3}{5}$ = 0.______

10. 12 × 5 = ______

11. Draw to show a 270° clockwise rotation on •.

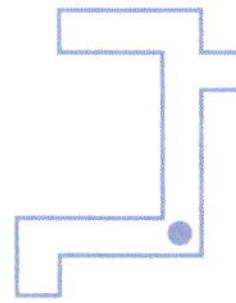
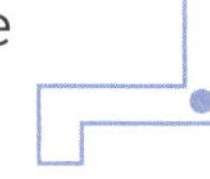

12. 6^2 = ______

13. (1000 ÷ 10) ÷ 5 = ______

14. $7\frac{2}{10} - \frac{9}{10}$ = ______

15. If there are 100 teachers and $\frac{1}{4}$ of them teach maths, what number do not teach maths? ______

Week 24

Wednesday

1. Parallelogram area = $b \times h$ = ________cm × ________cm = ________cm^2

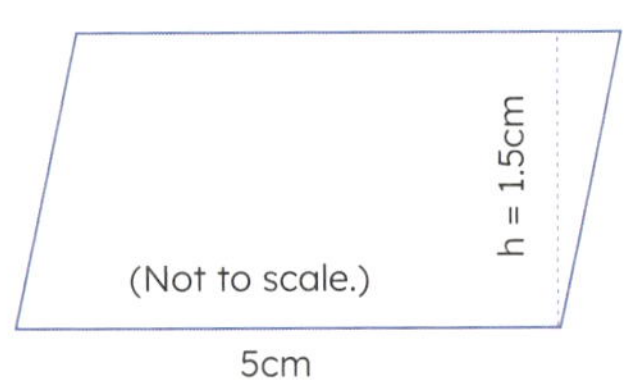

2. 35% of 620 = ________

3. 112 113 – 25 = ________

4. 82 × 68 = the sum of which column? ________

A	B	C
80 × 2 60 × 8 2 × 60 80 × 8	80 × 60 2 × 60 80 × 8 2 × 8	82 × 60 82 × 8 68 × 8 68 × 2

5. (200 ÷ 4) ÷ 10 = ________

6. Halve $\frac{1}{2}$. ________

7. $8\frac{2}{3}$ = ________.________

8. Double 13.5. ________

9. Area of triangle = ________cm^2

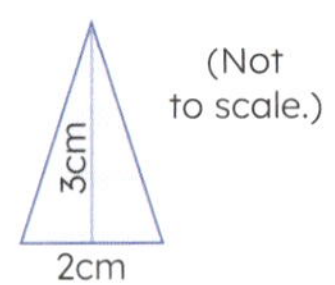

10. What is the volume of a cube if its sides are each 3m in length?

____

11. $\frac{4}{5}$ = 0.________

12. 9005mL = ________L

13. A farm has a square paddock with 6 fence posts along each side. How many fence posts are there altogether?

____

14. $y°$ = ________

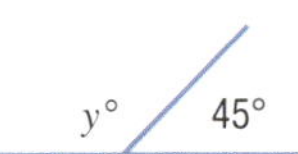

15. Grape juice costs $2.00 per 200mL. How much does 1L cost?

____

Thursday

1. Parallelogram area = ________cm^2

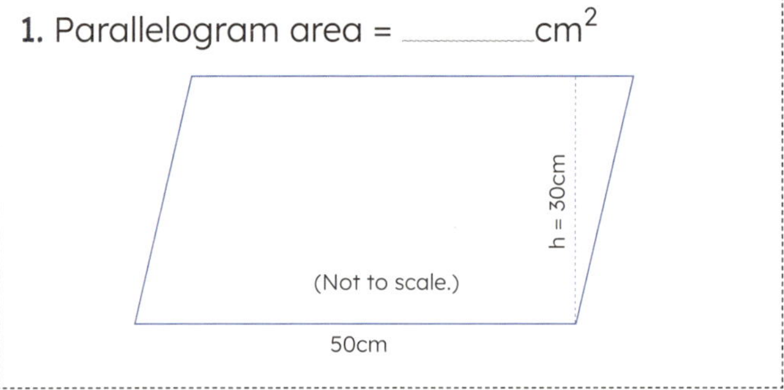

2. If it is 10:40 am Wednesday, what will be the day and time $24\frac{1}{2}$ hours later?

____

3. Orla's super-fast race car uses 3L of fuel per 5km. After 60km, how many litres will have been consumed?

____

4. $x°$ = ________

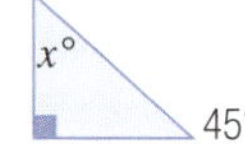

5. Round 2.565 to the nearest hundredth.

____

6. Which is the cheapest race horse to buy?

(a) Silver 25% off $2000 ☐

(b) Mr Ed 20% off $1800 ☐

(c) Phar Laptop 10% off $1900 ☐

7. $8\frac{2}{3} + 1\frac{2}{3}$ = ________

8. Double 60.5. ________

9. 15 + 25 + 35 + 55 = ________

10. Simplify 9:12. ________

11. 4035m = ________km

12. Draw the top view.

13. $40 - \frac{1}{2}$ = ________

14. Which two fractions are between 0.5 and 0.8?

(a) $\frac{2}{3}$ ☐ (b) $\frac{3}{4}$ ☐ (c) $\frac{2}{5}$ ☐

(d) $\frac{1}{4}$ ☐ (e) $\frac{1}{8}$ ☐

15. Flip two coins. The four outcomes are HH, ________, ________, or ________.

Problem-solving

Daku is paving a patio with these parallelogram-shaped slabs.

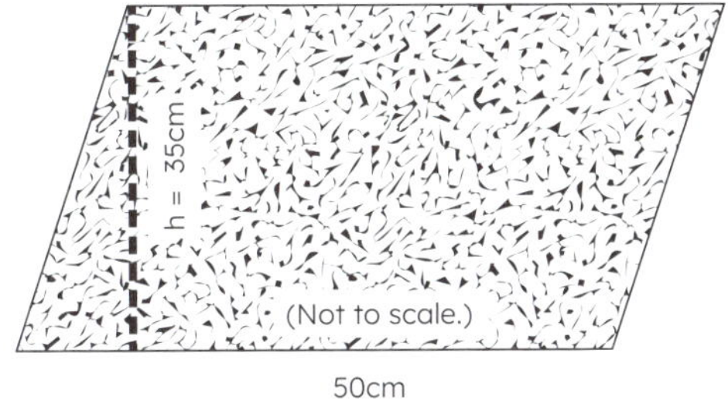

Daku has used 12 slabs so far.

If the patio has a total area of 49 000cm², how many more slabs does Daku need to lay? (Hint: Find out the area of the patio that has been covered first.)

Read the question again. **Think** about the information. Underline the important words.

Tick the strategy you will use to work out the answer:

- estimate and check ☐
- look for patterns ☐
- draw a diagram or picture ☐
- construct a table or graph ☐
- use materials ☐
- use a formula ☐
- something else. ☐

Solve it:

Reflect on the question and answer.

Check it. Circle another strategy on the list to work it out.

Show it:

Friday Review

Week 24

1. Parallelogram area = $b \times h$ =

_____ cm × _____ cm =

_____ cm²

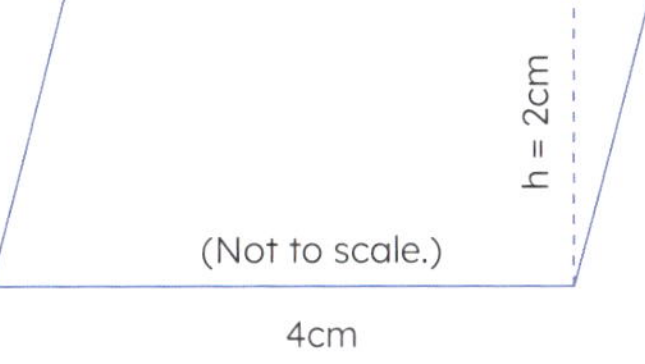

_____

2. $\frac{1}{4}$ = 0._____

3. Double this shape.

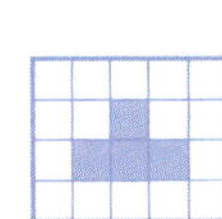

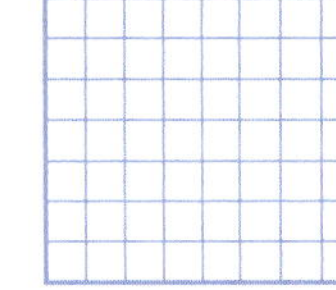

4. If $10 - e = 9\frac{1}{2}$, then

e = _____.

5. $7\frac{2}{5} - \frac{4}{5}$ = _____

6. (2000 ÷ 20) ÷ 5

= _____

7. 7^2 = _____

8. Area = _____ cm²

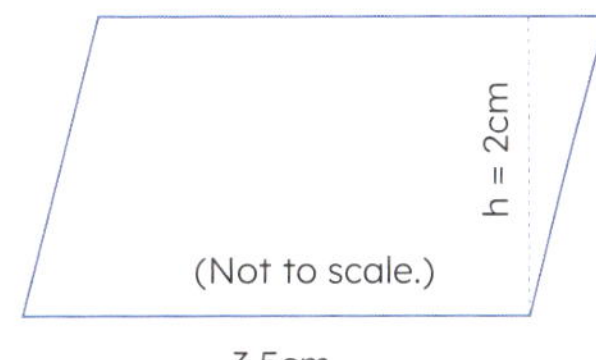

9. 3679L = _____ kL

10. $x°$ = _____

$x°$ 125°

11. 45 + 75 + 55

= _____

12. What is the cost of buying 2kg of olives if the price is $1.50 per 100g?

_____

13. If it is 7:45 pm, what will the time be 46 hours later?

_____

14. In a pie chart, graph the sports data: rugby 15%; lacrosse 25%; soccer 40%; golf 5%; tennis 10%; hockey 5%.

Favourite sports

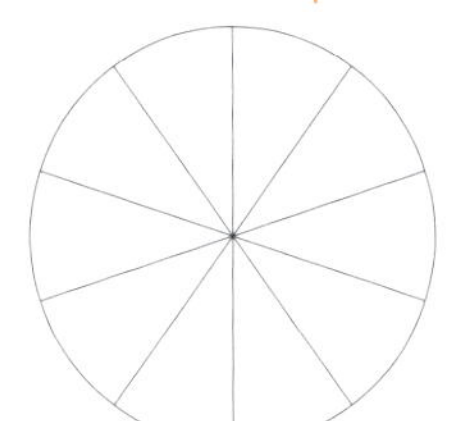

15. Measure the line AB. _____ mm

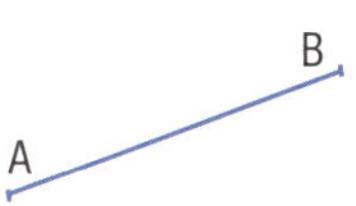

16. 8 × 19 × 50 = _____

17. Draw a 270° clockwise turn.

18. What is the probability of choosing a red bean from a jar containing 1 red and 99 green beans?

_____

Sp St

Monday

1. 10^1 = ______

2. If $3\frac{3}{4} + e = 4\frac{1}{2}$, then e = ______.

3. $a°$ = ______

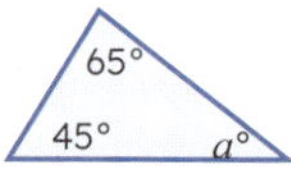

4. Continue the powers of two sequence:

   1, 2, 4, 8, ______, ______, ______

5. A water tank holds 6kL. It has leaked one-quarter of its capacity. How many litres have been lost? ______

6. If it is 6:00 am (ACST) in Adelaide, what is the time in Perth (AWST)?

   ______

7. $30 - \frac{1}{3}$ = ______

8. A green can weighs twice as much as a blue can. A yellow can weighs half of the green can. If the blue can weighs 4kg, the weight of the yellow can is

   ______kg and the green can is ______kg.

9. Write $\frac{5}{100}$ as a decimal. 0.______

10. What is the mean of the ages 13, 14, 13, 12, 19, 15, and 12?

    ______

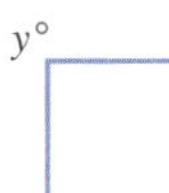

11. y = ______

12. Write in ascending order:

    0.2 × 5 0.11 8% 1 ÷ 2 0.9 × 0.1

    ______ ______ ______ ______ ______

13. 19 + 19 = ______ ÷ 2

14. Turn the rectangle 90° clockwise and draw the new position.

15. Your school canteen sells frozen yoghurt for 55c per 100g. What is the cost of a 300g purchase?

    ______

Tuesday

1. 10^3 = ______

2. 400 000, 600 000, 800 000, ______

3. If 900 – x = 200, then x = ______.

4. 16 + 16 = ______ ÷ 2

5. What do the internal angles to a triangle sum to?

   ______

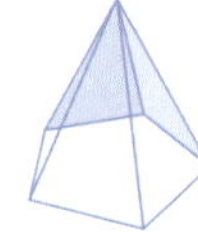

6. Name this 3D object.

   ______

7. How many faces does the above object have?

   ______

8. Round 649 749 to the nearest hundred thousand.

   ______

9. How many B boxes will fit evenly into box A?

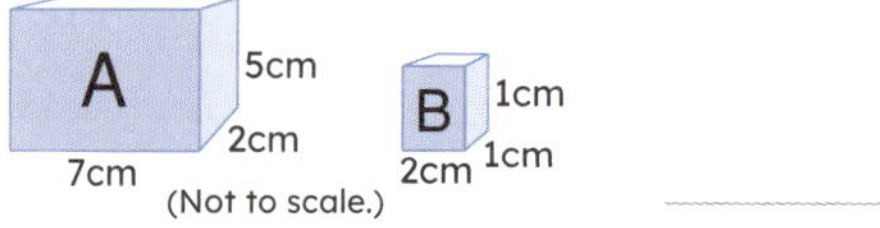

______

10. 250 000 + ______ = 1 000 000

11. How many sides does an icosahedron have? ______

12. 10 000g = ______kg

13. $10\overline{)854}$ = ______

14. Write 11:35 pm in 24-hour time. ______

15. You have $20.40. Of that, $2.40 is made of 20c coins and $10.00 is $1.00 coins. The rest is made of 50c coins. How many 50c coins do you have?

    ______

Wednesday

1. 10^2 = ______

2. Draw an irregular pentagon.

3. $6\overline{)7206}$ = ______

4. 603.85 ÷ 10 = ______

5. Turn the cylinder 180° anticlockwise and draw the new position.

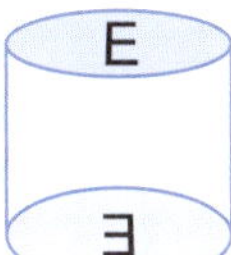

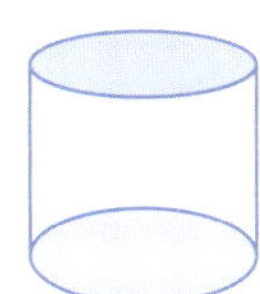

6. 15 + 14 = ______ ÷ 2

7. 250 000, 500 000, 750 000, ______

8. Halve $\frac{1}{5}$. ______

9. A train leaves a station at 0615. Its first stop is at 0631 and its next stop is at 0647. The next stop is 0703. What is the time interval between stops?

______

10. 600 + 300 = ______,

600 000 + 300 000 = ______

11. After folding this into a cube, which face is opposite D?

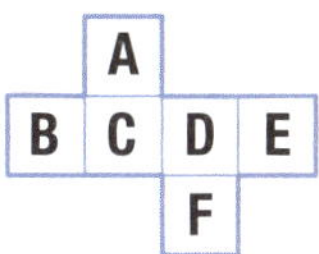

______

12. $1\frac{1}{4}$ = 1.______

13. y = ______

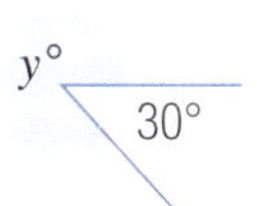

14. Which two shapes are shaded in half?

______ and ______

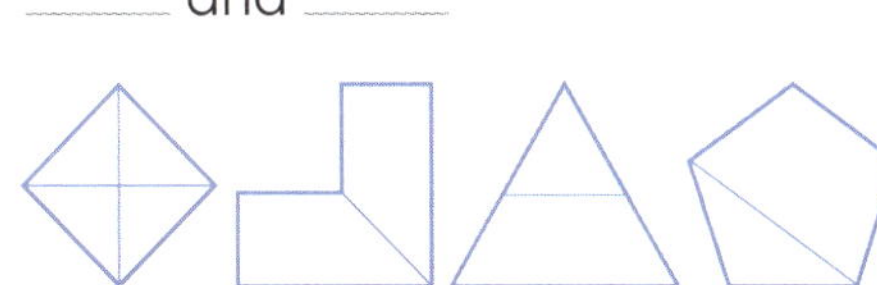

15. A pencil case holds 10 blue and five red pencils. What is the probability of choosing a red pencil?

______

Thursday

Week 25

1. 10^4 = ______

2. A truck leaves at 6:00 am, covers 600km, and arrives at its destination six hours later. How fast is the truck travelling?

______

3. A garden has 16 vegetables in a ratio of one turnip for every three carrots. How many turnips are in the garden?

______

4. 50 – 35 = 3 × ______

5. 40 – 30 ÷ 5 = ______

6. $8\frac{1}{3} + \frac{2}{3}$ = ______

7. Kylie ate toast for breakfast for four out seven days last week. How many times would you expect Kylie to eat toast in six weeks?

20 ☐ 24 ☐ 28 ☐

8. Does this object have rotational symmetry?

yes ☐ no ☐

9. 10 000, 9200, 8400, 7600, ______

10. Parallel means two lines in the same plane which never ______.

11. Write in descending order: $\frac{1}{2}$, 0.049, 0.1, 3%.

______ ______ ______ ______

12. 25 + 27 = ______ ÷ 2

13. Write the prime factors of 15.

______ and ______

14. 10 000mm = ______m

15. Match each capacity to its jar: 1.2L, 1150mL, 1.1L, 890mL.

b > d c < b b > a d < c a > c

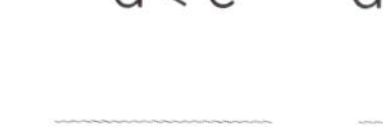

______ ______ ______ ______

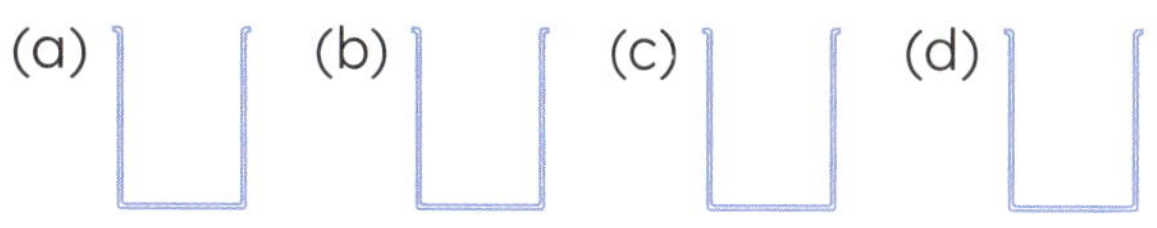

Problem-solving

Fill in the missing powers in this equation.

54 290 = (5 × 10☐) + (4 × 10☐) + (2 × 10☐) + (9 × 10☐)

What pattern do you notice in your answers?

Read the question again. **Think** about the information. Underline the important words.

Tick the strategy you will use to work out the answer:

- estimate and check ☐
- look for patterns ☐
- draw a diagram or picture ☐
- construct a table or graph ☐
- use materials ☐
- use a formula ☐
- something else. ☐

Solve it:

Reflect on the question and answer.

Check it. Circle another strategy on the list to work it out.

Show it:

Friday Review

1. 10^3 = ______
2. If 605 − x = 305, then x = ______.
3. Halve $\frac{1}{4}$. ______
4. 402.75 ÷ 10 = ______
5. An icosahedron has ten red faces and the rest are blue. What is the probability of rolling a red face? ______
6. Can parallel lines join? yes ☐ no ☐
7. What is the median of the ages 13, 14, 13, 12, 19, 15, and 12? ______
8. Does this object have rotational symmetry? yes ☐ no ☐

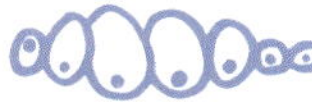

9. 10^5 = ______
10. (9000 ÷ 30) ÷ 3 = ______
11. What is the ratio of cars to bikes? ______

12. In Thursday question 3, what is the chance of randomly picking a turnip? ______
13. Write $2\frac{1}{4}$ as a decimal. ______
14. $a°$ = ______

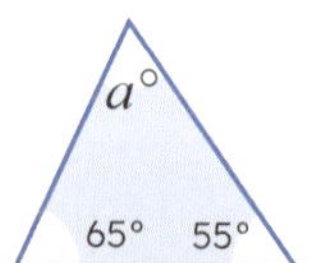

15. Rotate the cylinder 180° clockwise and draw the new position.

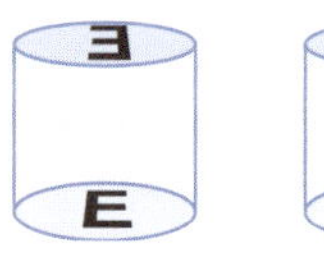

16. If $\frac{2}{3}$ of the class of 27 students ate cereal in the morning, how many students did not eat cereal? ______
17. How many faces does a square pyramid have? ______
18. Complete the pattern.

7		63
21	•	
	35	

▶

	56	
7	•	
21		35

▼

49		
	•	35
7	21	

◀

	•	

Week 26

Monday

1. $\frac{1}{2} \times \frac{1}{4}$ = ______

2. If 900 – e = 500, then e = ______.

3. 19 + 24 + 18 = ______

4. Write in descending order:
   5% 0.9 0.11 0.222

   ______ ______ ______ ______

5. Simplify $\frac{16}{20}$. ______

6. 60 × 50 = 25 × ______

7. Name this 3D object.
   (Not all 20 faces are shown.)

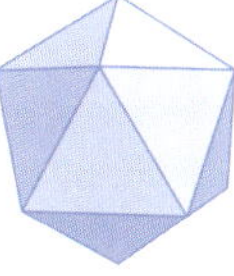

   ______

8. Halve 1300. ______

9. 0.8, 1.1, 1.4, 1.7, ______

10. 25% of $100.00 = ______

11. A road crew charges $5.50 per kilometre for painting a white road line. How much does 10km cost?

    ______

12. 7 × 8 = ______, 70 × 80 = ______

13. What is the time in Melbourne (AEST) if it is 6:30 pm in Darwin (ACST)?

    ______

14. Complete the garden plan.

    ☐ = $8m^2$

    $64m^2$ needs to be grass and $24m^2$ is tulips.

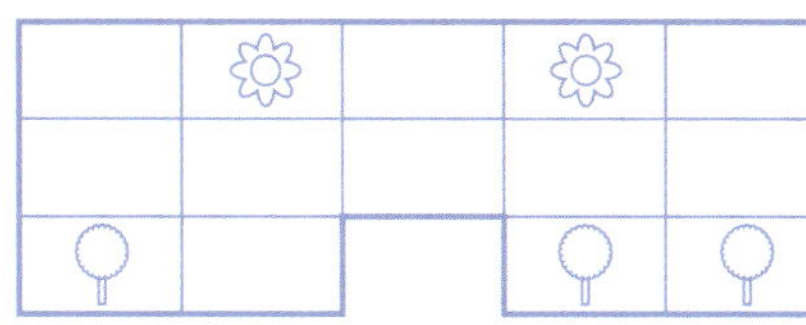

grass

tulips

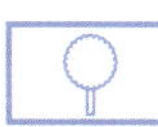

trees

15. 5 × 7 + 7 = ______

Tuesday

1. $\frac{3}{4} \times \frac{1}{5}$ = ______

2. What 3D object does the net make?

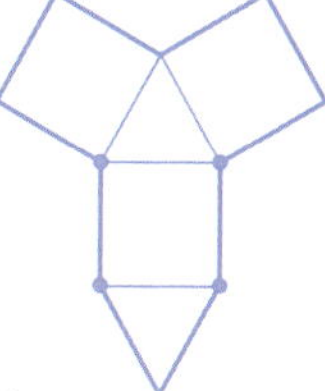

   ______

3. 100 000 – 68 000 = ______

4. Double 7500. ______

5. Draw an irregular hexagon.

6. 2500, 2250, 2000, ______

7. What is the ratio of boomerangs to stars?

   ______

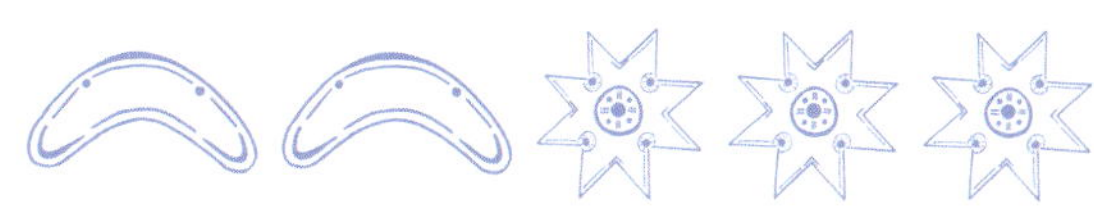

8. How many faces does a triangular prism have?

   ______

9. Name this 3D object.
   (Not all 12 faces are shown.)

   ______

10. Round 6805 to the nearest thousand.

    ______

11. If it is 1:30 am on Saturday, what will be the day and time 72 hours later?

    ______

12. 1 000 000 =

    300 000 + ______

13. 36 + 38 + 39 = ______

14. On leaving a town, a bus's odometer displayed 0481km.
    On its arrival to another town, the odometer read 0492km.
    Which two towns did the bus travel between?

    Burra 67km | Terowie 46km
    Hanson 56km | Yunta 59km

    ______

    ______

15. 300 ÷ 30 = ______

Week 26

Wednesday

1. $\frac{2}{3} \times \frac{1}{4} =$ ______

2. Which line is perpendicular to B? ______

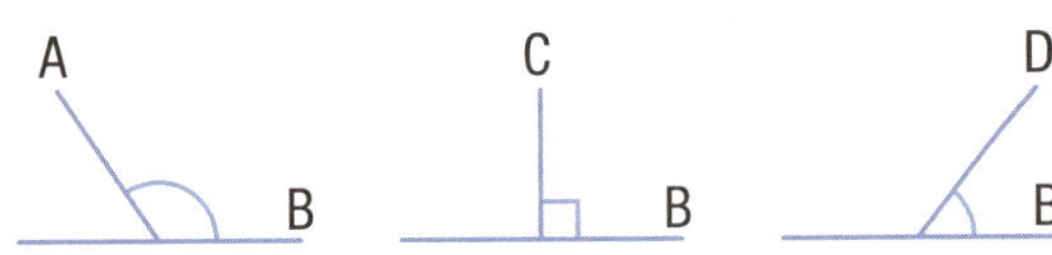

3. If it is 6:30 am on Friday, what will be the day and time $50\frac{1}{2}$ hours later?

______

4. $4^3 =$ ______ × ______ × ______ = ______

5. Order the numbers from most likely to least likely for the arrow to stop at when spun.

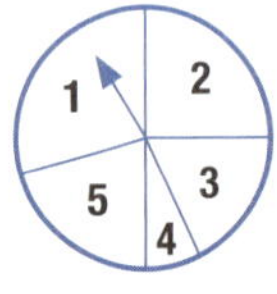

______ ______ ______ ______ ______

6. Continue the powers of three sequence:

1, 3, 9, 27, ______, ______, ______

7. If 10 × 5 = 50 and 7 × 5 = 35, then what is 17 × 5?

______

8. $y° =$ ______

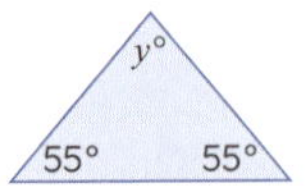

9. Write in descending order: 4%, $6\frac{1}{4}$, 0.5, 2.

______ ______ ______ ______

10. $\frac{12}{20} = 0.$______

11. 16 – 9 = ______, 1.6 – 0.9 = ______

12. (320 ÷ 80) × 9 = ______

13. $y° =$ ______

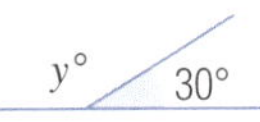

14. If you have $15.00 in $1.00 coins, $20.00 in 50c coins, and $10.00 in 20c coins, how many coins do you have altogether?

______

15. Write in ascending order: –4, 8, 3, –2, 0, –5.

______ ______ ______ ______ ______ ______

Thursday

1. $\frac{5}{6} \times \frac{1}{2} =$ ______

2. If a bus leaves its depot at 2310 and each stop is at 6 minute intervals, what time is the sixth and final stop?

______

3. Double 18 500. ______

4. If Matt walked from A to B, how many kilometres did Matt walk?

______

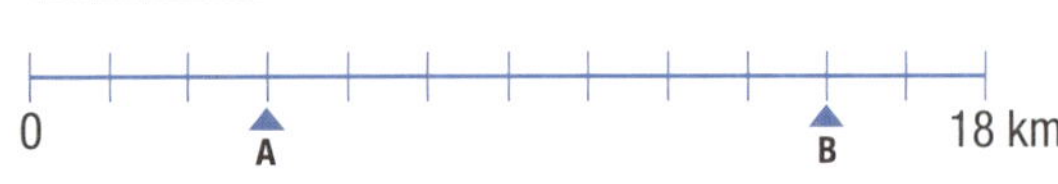

5. $7^2 =$ (a) $4^2 + 3^2$ ☐ (b) $10^2 - 51$ ☐

(c) $6^2 + 1^2$ ☐ (d) $5^2 + 2^2$ ☐

6. Halve 32 500. ______

7. Simplify $\frac{18}{21}$. ______

8. You have $40.00 of 50c coins. How many coins altogether?

______

9. Draw the rectangle after a 450° turn anticlockwise.

10. 14 + 9 = ______,

0.14 + 0.9 = ______

11. 12.65km = ______ m

12. 1 000 000 – 100 000 = ______

13. What is the difference between $1025 and $895? ______

14. 40% of $500 = ______

15. Complete the pattern.

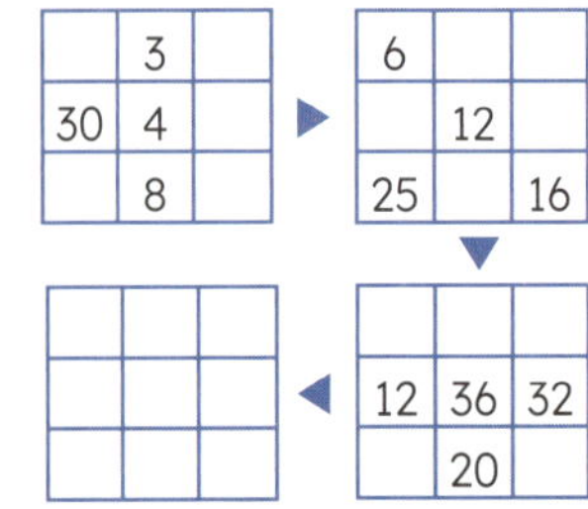

Problem-solving

Zoe went on a bug hunt in their garden. $\frac{5}{6}$ of the bugs they found had legs. $\frac{2}{3}$ of the bugs with legs were ladybugs.

What fraction of all the bugs Zoe found were ladybugs?

Read the question again. **Think** about the information. Underline the important words.

Tick the strategy you will use to work out the answer:

- estimate and check ☐
- look for patterns ☐
- draw a diagram or picture ☐
- construct a table or graph ☐
- use materials ☐
- use a formula ☐
- something else. ☐

Solve it:

Reflect on the question and answer.

Check it. Circle another strategy on the list to work it out.

Show it:

Friday Review

1. $\frac{1}{2} \times \frac{3}{4} =$ ________
2. Double 19 500.

 ____
3. If $64 = 2a$,

 then $a =$ ________.
4. If a bus leaves its depot at 1730 and stops at 8 minute intervals, what time is the fifth and final stop?

 ____
5. Draw a reflection.

6. $\frac{14}{20} = 0.$________
7. How many B boxes will fit evenly into box A?

 ____

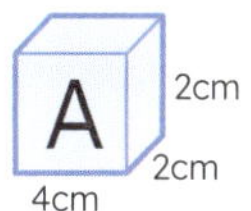

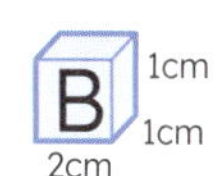

(Not to scale.)

8. Can a triangle, square, and circle tessellate together?

 yes ☐ no ☐
9. $\frac{2}{3} \times \frac{1}{6} =$ ________
10. 9.1m = ________mm
11. What is the ratio of footballs to netballs?

 ____

12. 30 × 40

 = 60 × ________

 = ________
13. 40% of $500.00

 = ________
14. Riki flipped two coins. They both landed on heads. The probability of this event is

 ______ in ______.
15. 4 × 7 − 2 = ________
16. Which is the acute angle?

 ____

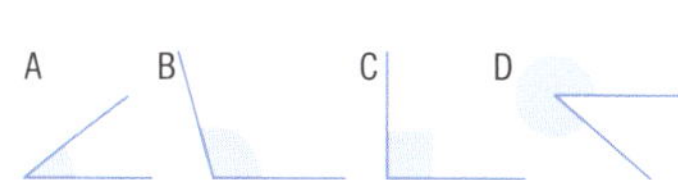

17. If it is 3:20 am on Monday, what will be the day and time in 49 hours?

 ____

 ____
18. If you are paid $4.50 per hour as a chocolate cake taste tester, what could you earn after five hours?

 ____

Week 26

Week 27

Monday

1. $-4 > 5$ true ☐ false ☐

2. What is the perimeter of a regular pentagon with 40mm sides?

____

3. Draw the top view.

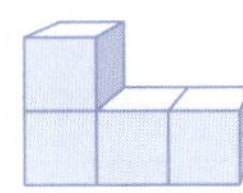

4. Write $10\frac{3}{8}$ as an improper fraction. _______

5. 350, 700, 1050, 1400, 1750, _______

6. If $9000 - e = 6000$, then $e =$ _______.

7. Start from 150 and count up by 250s. What is the fifth position in the sequence?

____

8. What is the angle at:

a? _______ °

b? _______ °

9. A 300g jar of blueberry jam has 40% fruit. How many grams of blueberries are there in the jar?

____

10. $40^2 + 2^2 =$ (_______ × _______) + (_______ × _______)

= _______ + _______

= _______

11. What is the mean of Don Bradboy's cricket scores of 80, 120, and 70?

____

12. Draw to show a $\frac{3}{4}$ turn anticlockwise.

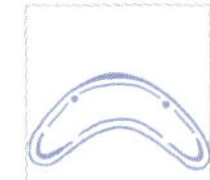

13. How many 20c coins make $6? _______

14. A regular pentagon has five internal angles that add to _______.

15. If you are paid $1\frac{1}{2}$ times the normal rate ($5.00 per hour), what do you earn for two hours' work?

____

Tuesday

1. $-8 > 3$ true ☐ false ☐

2. Alicia (a) has 3 times as many frogs (f) as Mia (m). Which expression can help show the number of frogs Alicia has?

(a) $f \times 3$ ☐ (b) $m \times 3$ ☐

(c) $a \times 3$ ☐ (d) $\frac{a}{3}$ ☐

3. $30^2 + 3^2 =$ (_______ × _______) + (_______ × _______)

= _______ + _______

= _______

4. If $2e = 32$, then $e =$ _______.

5. Write in ascending order: $\frac{3}{4}$, $\frac{4}{5}$, 1.2, 0.6.

_______ _______ _______ _______

6. Is 72 140 divisible by 5? yes ☐ no ☐

7. $2 + \frac{9}{100} + \frac{6}{10} =$ _______._______

8. A tap leaks at a rate of 0.4L per hour. How much water is wasted after $3\frac{1}{2}$ hours?

____

9. $y° =$ _______

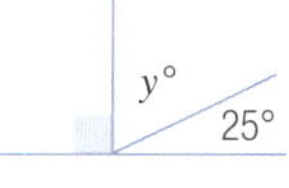

10. A chef has a $\frac{2}{3}$ measuring cup and a recipe that asks for six cups of flour. How many $\frac{2}{3}$ cups are needed?

____

11. How many hours are there between 4:30 pm on Monday and 4:30 pm on Tuesday?

____

12. $6 + 4 \times 10 =$ _______

13. $9 \times 3 =$ _______

14. Write 14.25 million as a numeral.

____

15. Shade 2.7mL.

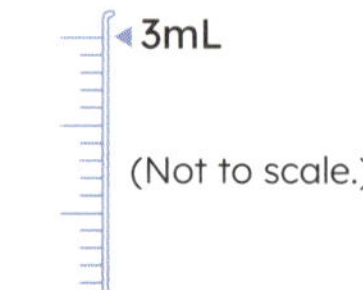

(Not to scale.)

Wednesday

1. –4 > 2 true ☐ false ☐

2. $\frac{8}{20}$ = 0.______

3. Luan collected donations from four classes in the school. Class 1 gave $21.50, class 2 gave $18.65, class 3 gave $22.15, and class 4 gave 35c. Which class was an anomaly?

______

4. Draw the left-side view.

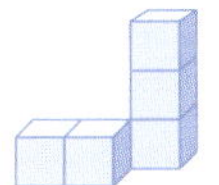

5. The biscuit tin has 45 biscuits in a ratio of four custard creams to one chocolate. How many custard creams are there?

______

6. $\frac{3}{4} - \frac{1}{2} = \frac{\square - \square}{4} = \frac{\square}{4}$

7. $x°$ = ______

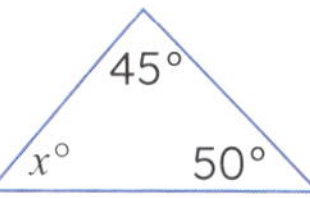

8. 4% = 0.______

9. Does this shape have rotational symmetry?

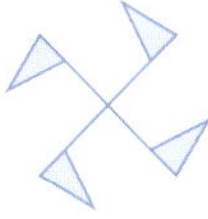

______

10. $\frac{2}{3}$ of 30 = ______

11. What number is halfway between 500 and 590?

12. $y°$ = ______

13. If you clean your room on Sunday and your mum pays you double time (normal rate is 35c per hour), how much would you earn after 3 hours?

______

14. What is the LCD for $\frac{3}{25}$ and $\frac{2}{25}$? ______

15. Measure $\overline{BD}$. ______mm

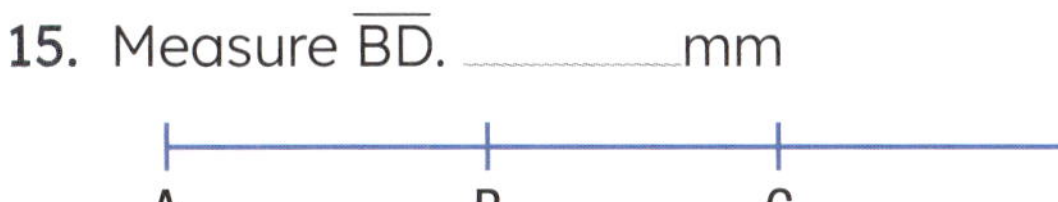

Thursday

1. –3 > –9 true ☐ false ☐

2. 400mm = 0.______m

3. Tick the name of this triangle.

(a) isosceles ☐

(b) scalene ☐

(c) even ☐

(d) square ☐

70°
55° 55°

4. What is the chance of guessing the correct answer for question 3?

______.

5. Halve 14.6. ______

6. 22 – 9 = ______

7. Place the horses in the correct order. Just Made It (A) could not pass Jockey Racer (B), Fast Legs Harry (C) was ahead of Jockey Racer (B), and Horse Power (D) tailed Just Made It (A).

1st ______ 2nd ______ 3rd ______ 4th ______

8. Double 14.6. ______

9. If $\frac{2}{3}$ of your teachers ride a bicycle to school, what fraction do not ride to school?

______

10. What is the LCD for $\frac{3}{25}$ and $\frac{4}{10}$? ______

11. Draw an irregular octagon.

12. Olivia is decorating the lounge room. One wall is 8m in length and 3m in height. A wallpaper roll covers 3m^2. How many rolls does Olivia need to cover that wall?

______

13. 11^2 = ______

14. What is the time difference between 3:15 pm on Friday and 3:30 pm on Saturday?

______

15. Write an example of a built structure that uses parallel lines.

______

Week 27

Week 27

Problem-solving

Look at the average temperatures in January for these cities.

City	Temp.	City	Temp.	City	Temp.
Bangkok	27°C	Jakarta	27.1°C	Seoul	-2.4°C
Beijing	-3.1°C	Karachi	18.1°C	Kathmandu	10°C
Hanoi	16.4°C	Tokyo	5.2°C	Ulaanbaatar	-21.6°C

Write a number sentence, using the < and > symbols, to describe the average January temperatures of the three cities in each column.

Then find the difference between the coldest and warmest temperatures for each column.

Read the question again. **Think** about the information. Underline the important words.

Tick the strategy you will use to work out the answer:

- estimate and check ☐
- look for patterns ☐
- draw a diagram or picture ☐
- construct a table or graph ☐
- use materials ☐
- use a formula ☐
- something else. ☐

Solve it:

Reflect on the question and answer.

Check it. Circle another strategy on the list to work it out.

Show it:

Friday Review

1 -5 > 2

true ☐ false ☐

2 450, 900, 1350, ________, 2250

3 Draw an irregular hexagon.

4 If 50 × 60 = a + 1800, then a = ________.

5 Banyu collected donations from three classes in the school. Class 1 gave $3.50, class 2 gave $95.95, and class 3 gave $83.50. Which class was an anomaly?

____

6 What is the mode of the Perth Viper's football scores of 35, 70, 90, and 35?

____

7 6.3 – 0.8 = ________

8 Shade 2.4mL.

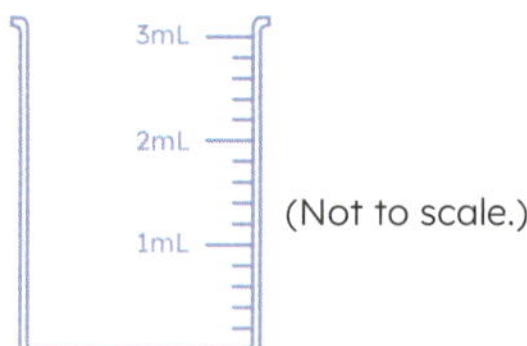

(Not to scale.)

9 What is the floor area of a 9m by 7m classroom?

____

10 $\frac{7}{20}$ = 0.________

11 2% = 0.________

12 Draw a top view.

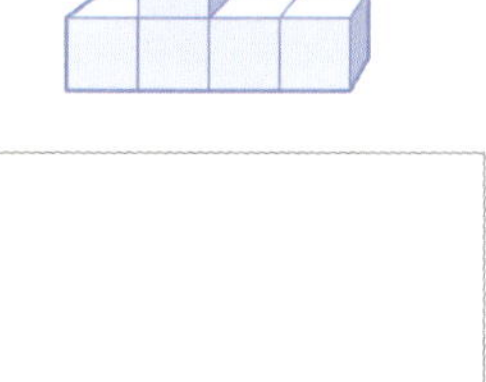

13 The formula for the area of a circle is

πr ________.

14 In Wednesday's biscuit tin (Q5), what is the chance of randomly picking a chocolate biscuit?

____

15 Draw to show as a $\frac{1}{4}$ turn anticlockwise.

16 Your cake taste testing job pays time and a half on Saturdays. If you work for two hours, what should you earn if the normal rate is $5.00 an hour?

____

17 What is the angle at:

A? ________

B? ________

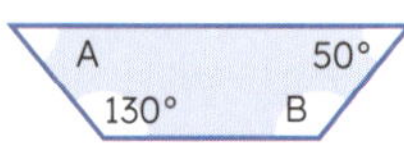

18 The cutlery drawer has 72 objects in a ratio of six spoons to six forks. How many forks are there?

____

Monday

1. Draw the line of reflectional symmetry onto the pattern.

2. $\frac{1}{4} \times \frac{1}{2} =$ ________

3. 5000, 4000, 3000, ________

4. Follow the instructions to label the cans: 4g, 40g, 400g, 4000g.

   A = $\frac{1}{10}$ of 400g B = 4kg

   C = 0.004kg D = 10% of 4kg

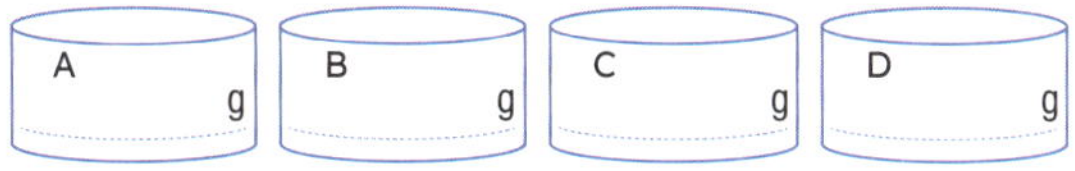

5. If 2000 = 400 + y, then y = ________.

6. 68 + 54 + 39 = ________

7. The prime factors of 35 = ________

8. 65mL = 0.________L

9. −3 > 8 true ☐ false ☐

10. $2\frac{1}{2}$ = ________%

11. What is the volume of a box which measures 20cm by 30cm by 40cm?

    ________cm^3

12. Draw to show a $\frac{1}{2}$ turn.

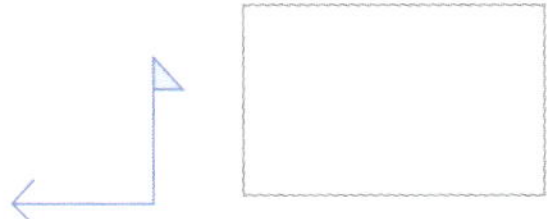

13. What is the mode of the class results:

    17%, 85%, 60%, 45%, 60%, 87%?

    ____

14. What is the range of the results? ________

15. What is the cost of buying 500g of cheese at $3.50 per 100g?

____

Tuesday

1. Draw the line of reflectional symmetry onto the pattern.

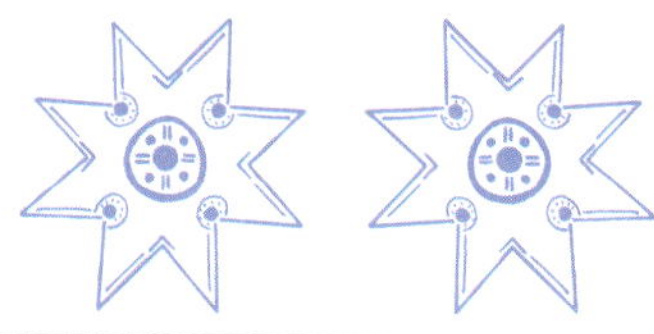

2. What is the LCD for $\frac{2}{3}$ and $\frac{6}{8}$? ________

3. (40 ÷ 8) × (8 − 5) = ________

4. If 1200 − a = 700, then a = ________.

5. Was $300, now ________.

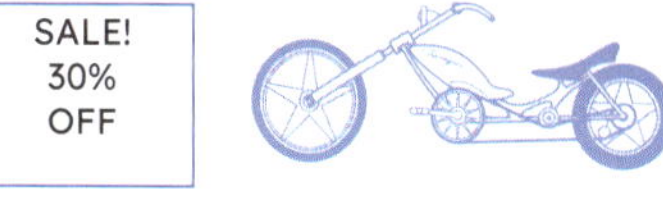

6. 89m = 0.________km

7. Write the two prime factors of 21.

   ________ ________

8. How many lines of symmetry does this letter shape have?

   ____

Mon	Tues	Wed	Thurs	Fri
4mm	21mm	23mm	19mm	22mm

9. Which amount of rainfall is the anomaly?

   ____

10. What is the range of the rainfall data?

    ____

11. 10^4 = ________

12. An obtuse angle is greater than ________° and less than ________°.

13. What is the product of 6 and 9? ________

14. −2 < 3 true ☐ false ☐

15. Name this type of triangle.

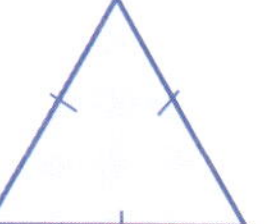

____

Week 28

Week 28

Wednesday

1. Does the boomerang symbol have reflectional symmetry?

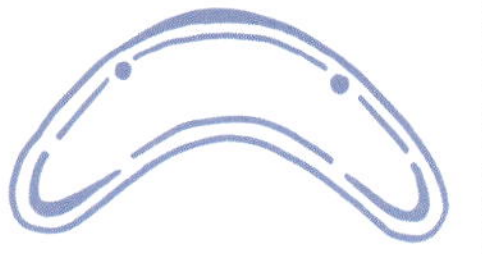

yes ☐ no ☐ nearly ☐

2. If $9\frac{1}{3} - e = 8\frac{2}{3}$, then e = ________.

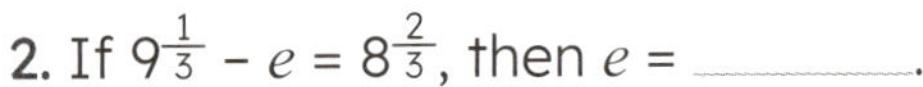

3. A reflex angle is greater than ________° and less than ________°.

4. $y°$ = ________

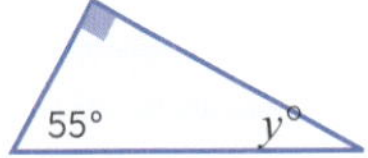

5. 97 + 64 + 33 = ________

6. How many hours are there between 5:00 am on Tuesday and 6:00 am on Thursday?

____

7. You ride 8km in a quarter of an hour. If you maintain that speed, how far could you ride in 2 hours?

____

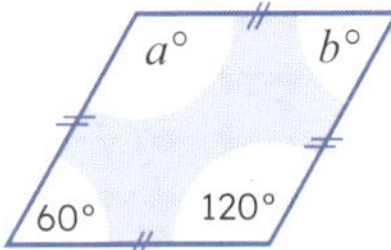

8. The internal angles of this ________ add up to ________°.

9. For the above shape $a°$ = ________° and $b°$ = ________°.

10. Add $9\frac{1}{2}$ hours to the time.

11. 300 000 + 800 000 = ________

12. 100 000 = ________ × ________ × ________ × ________ × ________ = 10^5

13. What are the prime factors of 24?

____

14. If petrol costs $1.05 per litre, how much is 10L? ________

15. Show a $\frac{1}{4}$ turn anticlockwise.

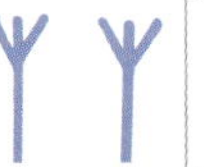

Thursday

1. Does the witchetty grub symbol have reflectional symmetry?

yes ☐ no ☐

2. What is the probability of randomly choosing a picture card from a deck of 52 playing cards?

____

3. What is the area of a 7m by 4m wall?

____

4. The formula for the area of a circle is ________.

5. An acute angle is less than ________°.

6. $\frac{4}{5}$ = 0.________ = ________%

7. What is the size of:

$x°$? ________

$y°$? ________

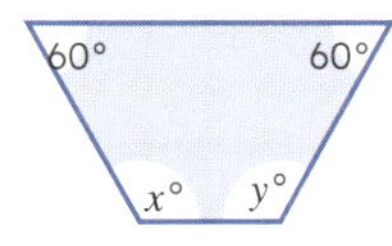

8. If 45 ÷ 5 = 3 × a, then a = ________.

9. Round 7.097 to the nearest tenth. ________

10. Rotate this shape 270° clockwise.

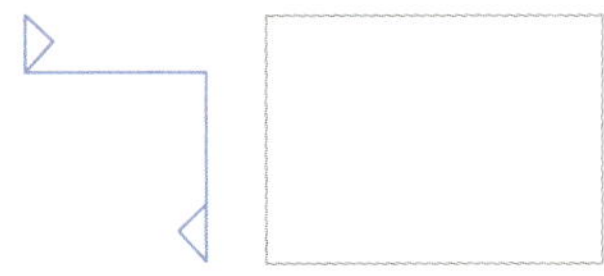

11. $100.00 – $30.00 = ________

12. 22 × 5 = ________

13. 3.5% = 0.________

14. 520L = 0.________kL

15. Correctly match each of the distances: 8m, 80m, 800m, 8000m.

A = 8km B = 0.008km

C = 10% of 8km D = $\frac{1}{10}$ of 800m

A = ________m B = ________m

C = ________m D = ________m

Problem-solving

Look at this collection of symbols used to represent common natural features, places, or objects in First Nations Australians' artworks.

boomerang | wombat tracks | meeting place | star | witchetty grub

Sort the symbols according to whether they have no symmetry, only reflectional symmetry, only rotational symmetry, or both reflectional and rotational symmetry.

Read the question again. **Think** about the information. Underline the important words.

Tick the strategy you will use to work out the answer:

- estimate and check ☐
- look for patterns ☐
- draw a diagram or picture ☐
- construct a table or graph ☐
- use materials ☐
- use a formula ☐
- something else. ☐

Solve it:

Reflect on the question and answer.

Check it. Circle another strategy on the list to work it out.

Show it:

Friday Review

1. What is the median of Tuesday's rainfall data? ______
2. Write three capital letters which are symmetrical. ______ ______ ______
3. Joe had cards with a prime factor of 35 and 15 written on each. After shuffling the cards, what is the probability of choosing a 5? ______
4. How many hours between 4:30 pm on Friday and 4:00 pm on Sunday? ______
5. Round 0.067 to the nearest hundredth. ______
6. The product of 9 and 8 is ______.
7. $\frac{3}{5}$ = 0.______ = ______%
8. How many lines of symmetry does this letter shape have? ______

9. Show as a $\frac{3}{4}$ turn clockwise.

10. $3\frac{1}{2}$ = ______%
11. The prime factors of 21 are ______ and ______.
12. 9000, 18 000, 27 000, ______
13. What is the mean of these three cricket scores: 30, 15, 0? ______
14. What is the perimeter of a 7cm regular pentagon? ______
15. If there's 40% off a $200.00 item, how much do you save? ______
16. Label the following masses: 9g, 90g, 900g, 9000g.

A = $\frac{1}{1000}$ of 9kg

B = 9kg

C = 10% of 9kg

D = $\frac{1}{10}$ of 900g

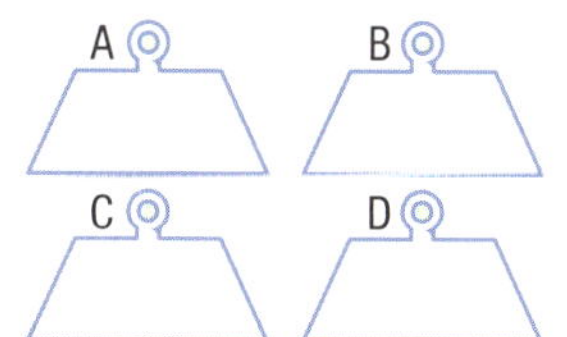

17. 40^2 = ______
18. What is the probability of randomly choosing a queen from a deck of 52 playing cards? ______

N A M Sp St P

Week 29

Monday

1. A lift started on floor –2 and went up five floors. On what floor did it stop?

   ______

2. 29 + 17 + 14 = ______

3. Name a 2D shape formed when you join two isosceles trapeziums together along their longest sides.

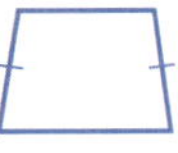

   ______

4. The value of 5 in 7.095 is ______.

5. If 10 × a = 5000, then a = ______.

6. Record 0.06 on the number line.

7. Write in ascending order: 0.99, $3\frac{2}{3}$, 5%.

   ______ ______ ______

8. 3500, 7000, 10 500, ______

9. 25 animals came to the waterhole on Monday. How many animals would you expect to come for a drink during the entire week?

   ☐ 175 ☐ 225 ☐ 250

10. 8 × 5 + 3 = ______

11. 0.86 × ______ = 860

12. Was $180, now ______!

FOR SALE
10%
OFF

13. $60\overline{)30\,000}$ = ______

14. For a car race qualifying heat, Anne achieved a finishing time of 2 min. 7.25 secs. If the minimum qualifying time is 2 min. 10 secs, by how much time did Anne qualify?

   ______

15. 2g = 0.______kg

Tuesday

1. A lift started on floor –1 and went up three floors. On what floor did it stop?

   ______

2. Fill in the missing number.

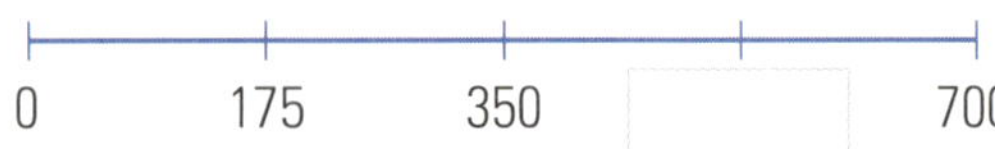

3. 25% of 6m = 50% of ______m

4. 650, 1050, 1450, 1850, ______

5. What is the cost of buying 40L of unleaded petrol at $1.10 per litre?

   ______

6. 300 – 33 = ______

7. If 60 × a = 3000, then a = ______.

8. Name this 2D shape.

   ______

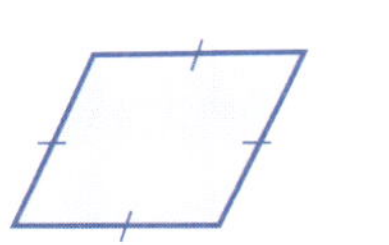

9. 4.9% = 0.______

10. $\frac{3}{4} \times \frac{1}{5}$ = ______

11. Bindi had a set of cards that all had a prime factor of 6 and 12 written on them. Record the numbers.

    ______, ______, and ______,

    ______, ______.

12. In your pocket, you find $10.00 of 50c coins and $4.80 of 20c coins.

    How many coins in total? ______

13. $6\overline{)84}$ = ______

14. 10^2 = ______ × ______ = ______

15. 14 × 5 = 10 × ______

Wednesday

1. A lift started on floor –5 and went up 18 floors. On what floor did it stop?

______

2. A class of 24 students completed a combined total of 432 laps for a school lapathon. Which expression matches the average number of laps per person? (Tick your answer.)

(a) 432 + 24 ☐ (b) 432 – 24 ☐

(c) 432 ÷ 24 ☐ (d) 24 × 432 ☐

3. $\frac{5}{20} > \frac{1}{2}$ true ☐ false ☐

4. 149.7 ÷ 100 = ______

5. Double $\frac{1}{4}$. ______

6. Which two prime numbers, when added, equal 24?

______

7. If 9.3 – a = 8.9, then a = ______.

8. If chocolate chip ice cream costs $4.50 for 1L, how much does 5L cost?

______

9. $\frac{8}{10} - \frac{2}{5}$ = ______

10. A jam factory loses $\frac{1}{10}$ of its 1000 jars due to breakage. How many are lost?

______

11. 25% of 8m = 50% of ______m

12. Write the missing numbers.

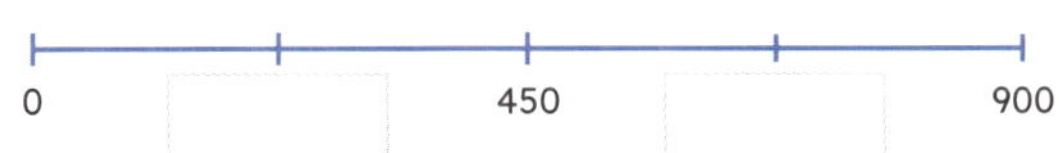

13. 40 + 60 + 90 = ______

14. How many hours between 6:30 pm on Thursday and 7:00 am on Saturday?

______

15. The Horse Hotel charges $8 per horse per day for its stables. If you had four horses for the month of August, what would the total cost be?

______

Thursday

Week 29

1. A lift started on floor -2 and went up 24 floors. On what floor did it stop?

______

2. A blue train leaves the station at the time of 0711 and the next station is 43 minutes away. What time does it arrive?

______

3. $y°$ = ______

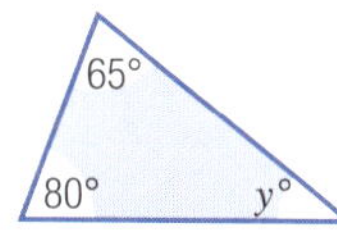

4. If 40 × a = half of 400, then a = ______.

5. Write the missing numbers.

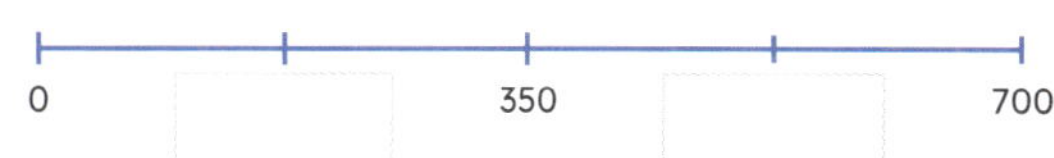

6. Double $15\frac{1}{2}$. ______

7. Halve 390. ______

8. 232.3 ÷ 100 = ______

9. Round 16 602 to the nearest thousand. ______

10. 2.993, 2.996, 2.999, ______

11. Draw the reflection.

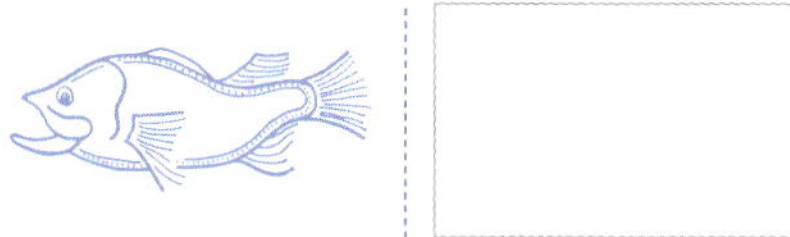

12. What is the total mass of these three chocolate boxes? ______

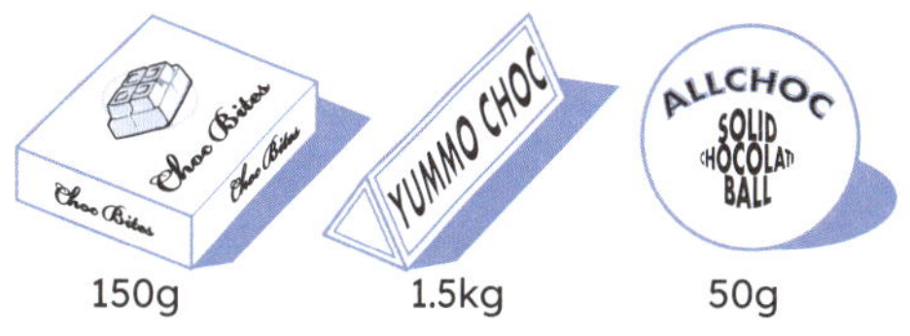

13. 7mm = 0.______m

14. 12.5%, 25%, 37.5%, ______, ______

15. $y°$ = ______

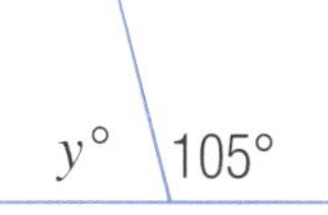

Week 29

Problem-solving

Luna was trialling a new form of transportation. They started off at sea level, then flew to the top of Mount Kosciuszko (2228m above sea level), dived down to the Titanic (3800m below sea level), visited the Dead Sea (414m below sea level), flew to the tip of Mount Everest (8849m above sea level), dived down the Mariana Trench (10 984m below sea level), and returned to sea level.

Across the entire trip, what was Luna's total change in altitude?

Read the question again. **Think** about the information. Underline the important words.

Tick the strategy you will use to work out the answer:

- estimate and check ☐
- look for patterns ☐
- draw a diagram or picture ☐
- construct a table or graph ☐
- use materials ☐
- use a formula ☐
- something else. ☐

Solve it:

Reflect on the question and answer.

Check it. Circle another strategy on the list to work it out.

Show it:

Friday Review

1. A lift started on floor −1 and went up eight floors. On what floor did it stop?

 ______

2. Which expression matches? What is the total mass of 25 packets of 125g chocolate?

 (a) 25 × 125 ☐

 (b) 25 + 125 ☐

 (c) 125 − 25 ☐

 (d) $\frac{125}{25}$ ☐

3. 5mm = 0.______m

4. Fill in the missing number.

 0 ______ 550

5. 25% of 10m

 = 50% of ______m

6. y° = ______

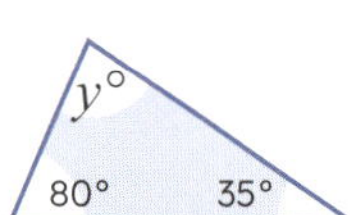

7. The value of 7 in 1 769 302 is

 ______.

8. $\frac{1}{4} \times \frac{2}{5}$ = ______

9. A lift started on floor −3 and went up 39 floors. On what floor did it stop?

 ______

10. If $75 = y \times 3$, then

 y = ______.

11. Name this shape. ______

12. Round 71 090 to the nearest thousand.

 ______

13. Write in descending order: 2.34, 18%, 0.11, 90%.

 ______

14. $\frac{1}{2} > \frac{4}{20}$

 true ☐ false ☐

15. What is the area of a 2.4m by 0.5m table?

 ______m^2

16. Brendan received $5.50 in pocket money each week. How much money would you expect Brendan to receive in eight weeks?

 (a) $38.59 ☐

 (b) $41.25 ☐

 (c) $44.00 ☐

17. Which type of prism has 7 faces?

 ______

18. A strawberry jam factory loses $\frac{1}{10}$ of its product to hungry workers. If it produces 1000L per week, how much does it lose each week?

 ______L

Sp

P

Monday

1. One floor of an apartment block has a volume of $150m^3$. If all the floors have the same volume, what is the volume of a five-storey apartment building?

   ________ m^3

2. Write $\frac{1}{8}$ as a decimal. ________

3. How many hours are there between 3 pm Friday and 8:30 pm Saturday?

   ____

4. If $a + 11.7 = 12.5$, then $a =$ ________.

5. Carlos bought a new guitar at a 25% discount. If the original price was $400, what did Carlos pay and how much was saved?

   ________ paid

   ________ saved

6. (200 ÷ 2) × (50 – 45) = ________

7. The ratio of adults to children watching an action movie is 5:1. If there are 100 adults, how many children are there?

   ____

8. If you swim 15 laps of a 50m pool, how many metres have you swum? ________

9. 500, 450, 400, ________

10. $5y = 25$

    $y =$ ________

11. 7 × 4 × 3 = ________

12. If a blue circle has a radius of 270cm, its diameter is

    ________cm.

13. If a painter charges $3.00 per square metre to paint a 7m × 2m wall, how much will it cost for one coat of paint?

    ____

14. 5 + 8 × 5 = ________

15. Halve 2090. ________

Tuesday

1. One floor of an apartment block has a volume of $220m^3$. If all the floors have the same volume, what is the volume of a five-storey apartment building?

   ________ m^3

2. Draw the line of reflectional symmetry onto the pattern.

3. What is the mean of the shoe sizes: 8, 5, 4, 8, 7, and 10? ________

4. If $y + 1200 = 3000$, then $y =$ ________.

5. $9y = 18$

   $y =$ ________

6. 24 + 6 ÷ 2 = ________

7. Write the missing number.

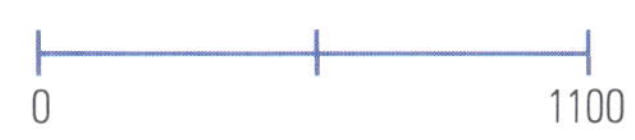

8. –4 > 2 ☐ true ☐ false

9. Round 339 487 to the nearest ten thousand.

   ____

10. 400 sausages were sizzled at school. Each student ate four. How many students were there? ________

11. Measure $\overline{XY}$. ________mm

    X ———————— Y

12. Mark on the line what the probability is of randomly selecting a blue marble from a hat if there are 15 blue marbles and five green marbles in it?

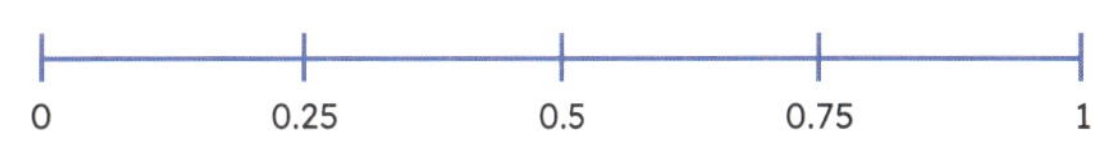

13. 110m = 0.________km

14. $9^2 =$ ________

15. $7 \times 10^5 =$ ________

Wednesday

1. What is the volume of an apartment of 10m by 7m by 2.5m?

 ________ m^3

2. If the apartment block has four of these sized apartments, what is the volume of the building?

 ________ m^3

3. If Reese walked from A to B, how many metres did they walk? ________

4. Double 0.95. ________

5. Circle the symmetrical letter: S E

6. $y°$ = ________

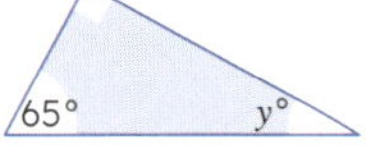

7. If 34 ÷ n = 6.8, then n = ________.

8. If a road builder charges $400.00 to create 100m of road, what is the cost of 1km?

 ____

9. Name this shape.

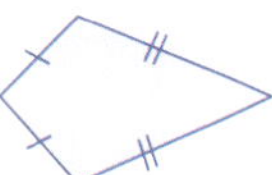

 ____

10. 0.2, 0.6, 1, 1.4, ________

11. Simplify the ratio 8:20. ________

12. Read the results to place the horses in the correct finishing order.

 Dan finished ahead of Jan. Tan was slower than Jan. Dan could not pass Han.

 1st – ________ 2nd – ________

 3rd – ________ 4th – ________

13. Jenny, a scientist, poured 0.4mL of H_2O into the jar. Shade the level.

 1mL

 (Not to scale.)

14. If 100 – e = 60, then e = ________.

15. 8×10^4 = ________

Thursday

1. What is the volume of an apartment of 12m by 8m by 2.5m?

 ________ m^3

2. If the apartment block has six of these sized apartments, what is the volume of the building?

 ________ m^3

3. $y°$ = ________

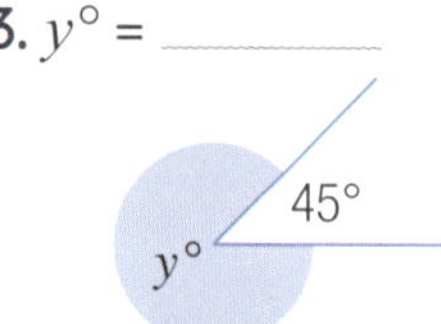

4. Is 617 divisible by 9? ________

5. 25% of 3kg = 50% of ________kg

6. Together, angles a and b equal ________°.

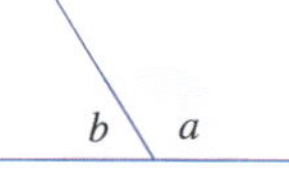

7. $\frac{1}{5} \times \frac{3}{4}$ = ________

8. What is the largest odd number that can be made from four of these digits: 7, 3, 5, 8, 3, 9?

 ____

9. An acute angle is > ________°

 and < ________°.

10. 4, 12, 20, 28, ________

11. How many B boxes will fit evenly into box A?

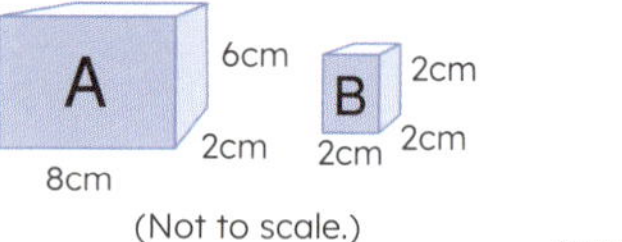

 (Not to scale.) ________

12. Round 10.056 to the nearest tenth. ________

13. Alicia paid $100 for a horse and a pony. Raif paid $170 for two horses and a pony. What is the cost of a horse?

 ____

14. 3×10^3 = ________

15. $3y = 12$

 y = ________

Problem-solving

Each of the first 12 floors of an apartment building has a volume of $400m^3$. Each of the next 12 floors has a volume of $350m^3$. Each of the final 12 floors has a volume of $300m^3$.

What is the total volume of the 36-storey apartment building?

Read the question again. **Think** about the information. Underline the important words.

Tick the strategy you will use to work out the answer:

- estimate and check ☐
- look for patterns ☐
- draw a diagram or picture ☐
- construct a table or graph ☐
- use materials ☐
- use a formula ☐
- something else. ☐

Solve it:

Reflect on the question and answer.

Check it. Circle another strategy on the list to work it out.

Show it:

Friday Review

1. What is the volume of an apartment of 10m by 8m by 2.5m?

   ______ m^3

2. If the same apartment block has eight of these sized apartments, what is the volume of the building?

   ______ m^3

3. Round 19 359 to the nearest 1000.

   ______

4. $3a = 24$

   $a =$ ______

5. Simplify 9:27. ______

6. If $\frac{1}{3}$ of $y = 10$, then

   $y =$ ______.

7. 2.2% = 0.______

8. Draw the line of reflectional symmetry onto the pattern.

9. What will the time be 12 hours from 4:45 am?

   ______

10. $x° =$ ______

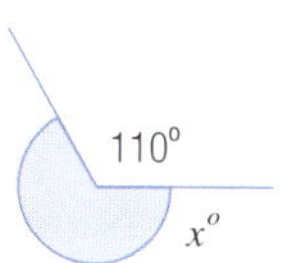

11. 9703m = ______ km

12. 25% of 2kg =

    50% of ______ kg

13. Flip two coins. The four outcomes are H and H,

    ______ and ______,

    ______ and ______,

    ______ and ______.

14. $\frac{2}{5}$ = 0.______

15. $-4 < 0$

    ☐ true ☐ false

16. 6.3, 6.9, 7.5, ______

17. What is the median of the shoe sizes: 8, 5, 4, 8, 7, 10?

    ______

18. If Tyrone swims 25 laps of a 50m pool, what distance have they covered?

    ______

Week 30

N A M Sp St P

Week 31

Monday

1. Shade the pie chart to show the favourite dog name survey results.

Mutt	10%
Spike	5%
Bella	25%
Spot	10%
Tom	30%
Mister	20%

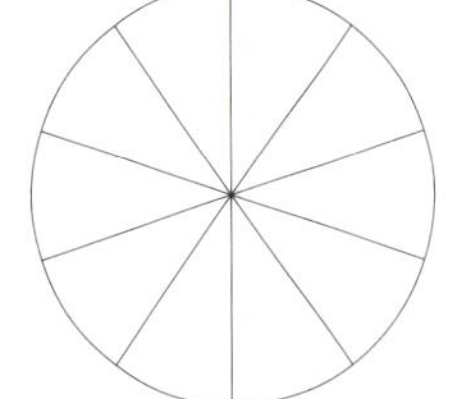

2. 0.96, 0.97, 0.98, 0.99, ________

3. What is the perimeter of a regular hexagon with 60mm sides?

____

4. Blue lolly (b) jars hold 300g more than red lolly (r) jars. Tick the expression that shows the mass of a red lolly jar.

(a) $b - 300$ ☐ (b) $r + 300$ ☐

(c) $b + 300$ ☐ (d) $300 \times r$ ☐

5. $50 \times 80 =$ ________

6. $\frac{1}{2} < \frac{1}{10}$ true ☐ false ☐

7. $18 + 7 + 12 =$ ________

8. Write in ascending order:

20% of 4, −3, $\frac{2}{3} \times 2$, 1.21, 42×0.1

____

9. A lift started on floor -1 and went up nine floors. On what floor did it stop?

____

10. Measure line $\overline{XY}$. ________mm

X Y

11. What is the perimeter?

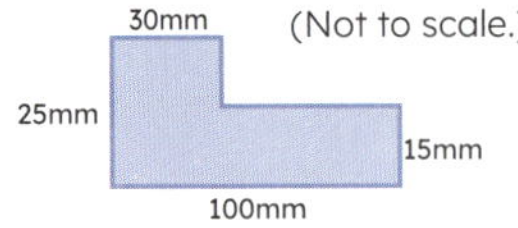

____

12. 6% = 0.________

13. A drink maker uses a ratio of one part fruit juice to four parts water. If it makes one litre of drink, how much fruit juice is used?

________mL

14. Which of these triangles has no lines of symmetry?

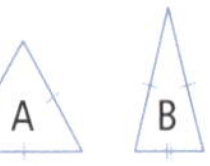

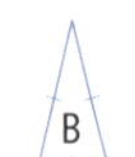

C

____

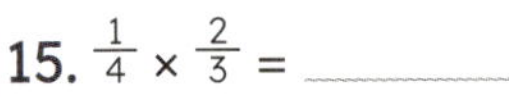

15. $\frac{1}{4} \times \frac{2}{3} =$ ________

Tuesday

1. Shade the pie chart to show the favourite animal survey results.

Koala	10%
Wombat	20%
Dingo	15%
Possum	20%
Quoll	20%
Bilby	15%

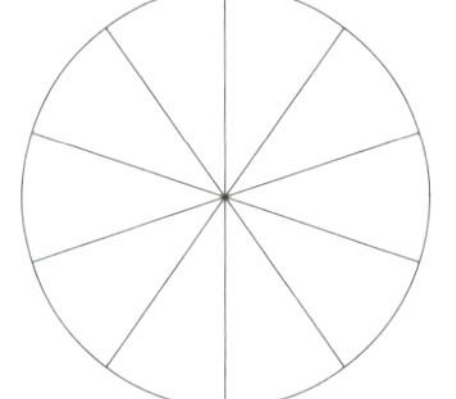

2. $\frac{3}{4} > \frac{1}{2}$ true ☐ false ☐

3. Draw the line of reflectional symmetry onto the pattern.

4. 75 000, 150 000, 225 000, ________

5. Simplify the ratio 8:16. ________

6. Write three capital letters that are symmetrical. ________ ________ ________

7. If a road builder charges $295 per 1km, how much does 3km cost?

____

8. The widest angle is ________, and is opposite the longest side, which is labelled ________.

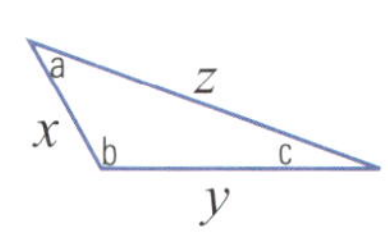

9. $6^2 =$ ________

10. Continue the powers of four sequence:

1, 4, 16, ________, ________, ________

11. Measure line $\overline{ABC}$. ________mm

A C B

12. A drink maker uses a ratio of one part fruit juice to three parts water. If it makes 1L of drink, how much water is needed?

________mL

13. How many edges does this hexagonal prism have?

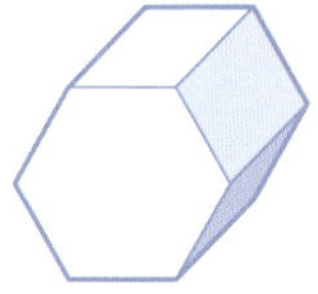

____

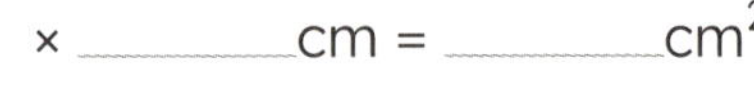

14. The product of 6 and 7 is ________.

15. Area = $b \times h$ = ________cm × ________cm = ________cm^2

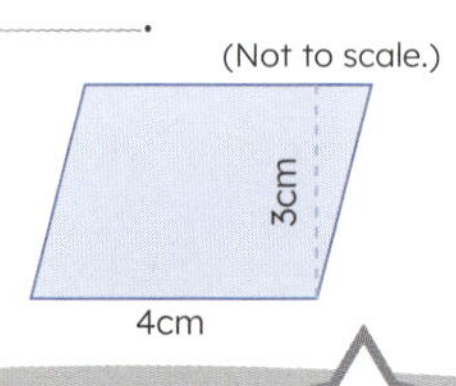

Wednesday

1. Shade the pie chart to show the favourite milkshake flavour survey results.

Vanilla	5%
Banana	30%
Berry	25%
Peach	20%
Melon	10%
Lemon	10%

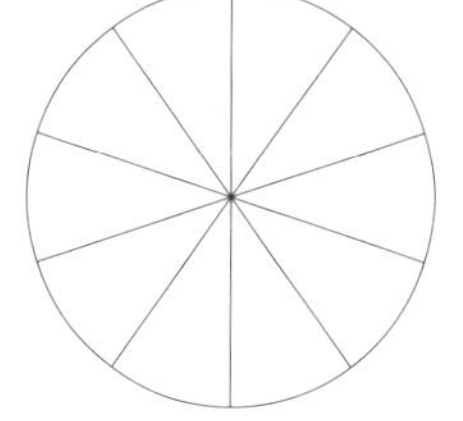

2. Write an expression to solve this problem.

   Ben, the local baker, bakes 40 loaves of wholemeal bread each Monday, Wednesday, and Friday. How many wholemeal loaves does Ben bake each week?

   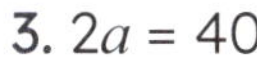

3. $2a = 40$

   a = ____

4. 1.07, 1.08, 1.09, ____
5. 60 000 + 40 000 = ____
6. If a road builder charges $8500 per km, how much will 5km cost?

   ____

7. 1mm = $\frac{1}{1000}$m = 0.____m
8. –9 < –4 true ☐ false ☐
9. Write 110% as a decimal. ____
10. $3\frac{1}{5}$ =

    (a) 3.15 ☐ (b) 3.2 ☐ (c) 3.5 ☐

11. 7×10^4 = ____
12. 7 + 6 × 5 = ____
13. $50\overline{)10\ 000}$ = ____
14. 600 × 500 = ____
15. Penny, a perfumer, made 28mL of petunia perfume. Shade this amount.

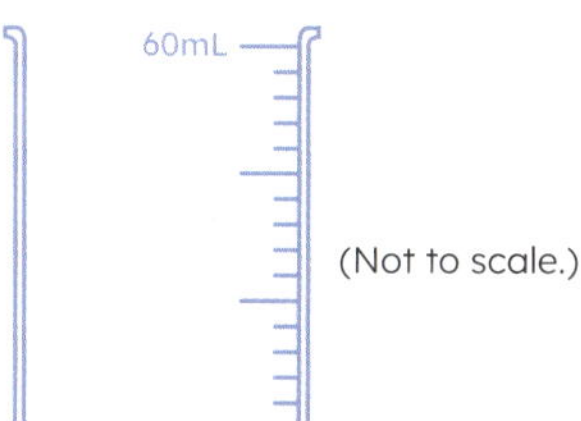

(Not to scale.)

Thursday

Week 31

1. Shade the pie chart to show the favourite dance survey results.

Tango	10%
Waltz	5%
Ballet	15%
Folk	15%
Salsa	35%
Swing	20%

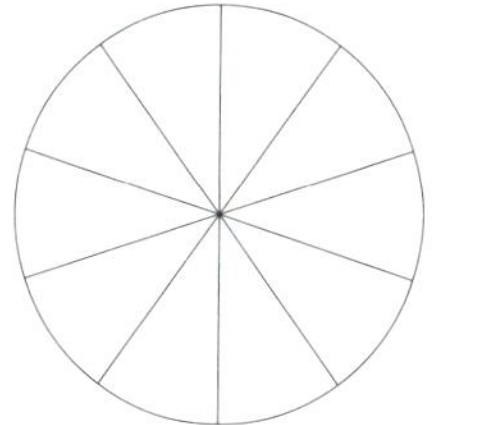

2. What is the time difference? ____

3. Turn this triangle 180° clockwise and draw the new position.

4. 50, 5, 100, 10, 200, 20, ____
5. 16 + 9 + 17 = ____
6. 900 × 500 = ____
7. A pancake recipe requires $1\frac{2}{3}$ cups of flour to four cups of milk. If eight cups of milk are used, how many cups of flour are needed?

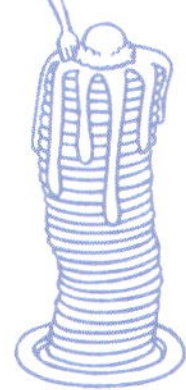

   ____

8. 5.5 – 0.9 = ____
9. Round 2.449 to the nearest whole. ____
10. $\frac{3}{5} < \frac{1}{2}$ true ☐ false ☐
11. 3.245, 3.145, 3.045, ____, ____
12. What is the probability of correctly guessing a heads during a one-coin flip?

    ____

13. Round 2.056 to the nearest tenth. ____
14. What is the sum of a quadrilateral's internal angles?

    ____

15. What is the ratio of 'V' to 'II' jerseys?

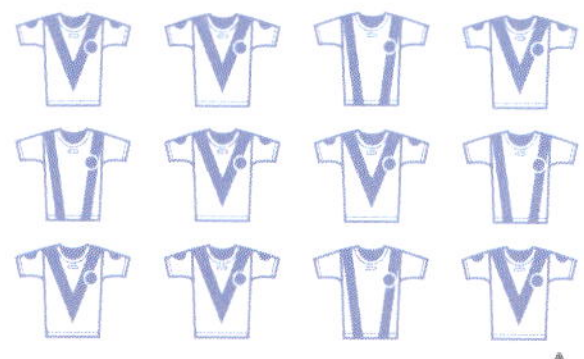

    ____

Problem-solving

The students in Class 7 measured their heights and plotted them onto a bar chart. The results were: 1.56m, 1.62m, 1.58m, 1.57m, 1.53m, 1.52m, 1.6m, 1.63m, 1.62m, 1.56m, 1.54m, 1.55m, 1.61m, 1.67m, 1.51m, 1.59m, 1.5m, 1.46m, 1.51m, 1.56m, 1.59m, 1.61m, 1.58m, and 1.54m.

Recreate the bar chart by drawing it or by using a digital program. Which height was the most common? Which two heights sit outside the others (outliers)?

Read the question again. **Think** about the information. Underline the important words.

Tick the strategy you will use to work out the answer:

- estimate and check ☐
- look for patterns ☐
- draw a diagram or picture ☐
- construct a table or graph ☐
- use materials ☐
- use a formula ☐
- something else. ☐

Solve it:

Reflect on the question and answer.

Check it. Circle another strategy on the list to work it out.

Show it:

Friday Review

1. Shade the pie chart to represent the survey's results.

Parrot	15%
Cockatoo	5%
Emu	10%
Kookaburra	20%
Galah	45%
Cassowary	5%

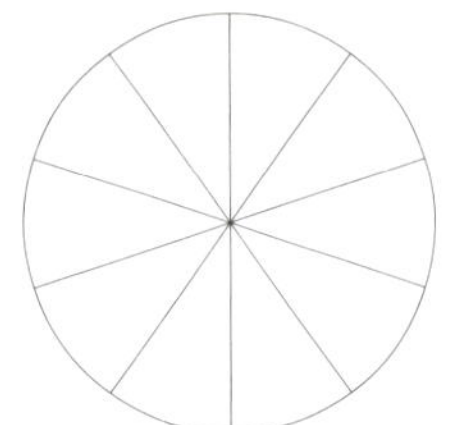

2. How many vertices does this object have?

______

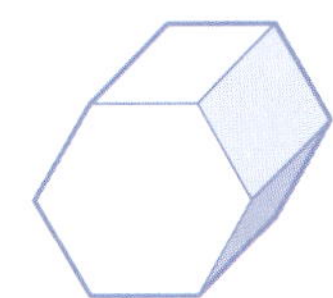

3. 30 000 – 4400

= ______

4. $\frac{1}{10} > \frac{1}{3}$

true ☐ false ☐

5. What is the probability of correctly guessing two heads during a two-coin flip?

______

6. 0.1 = ______%

7. 60 000, 150 000, 240 000, 330 000,

______

8. What is the time difference?

______

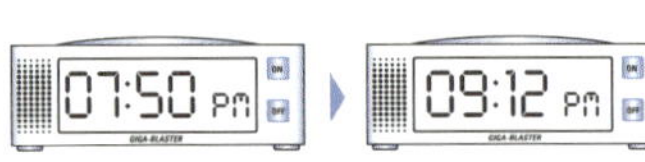

9. 4^2 = ______

10. $\frac{3}{4} \times \frac{7}{8}$ = ______

11. If a road contractor charges $7500 per km, how much would 5km cost?

______

12. Area = $b \times h$

= ______cm

× ______cm

= ______cm^2

(Not to scale.)

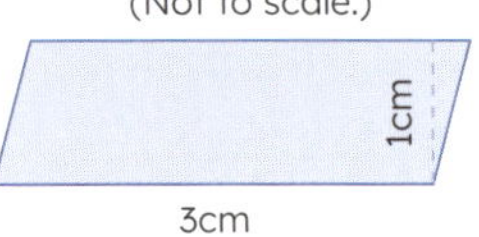

13. Share $260.00 equally among four people.

______

14. The product of 6 and 9 is ______.

15. 820 – 60 = ______

16. $3a = 30$

a = ______

17. Continue the powers of two sequence:

1, 2, 4, 8, ______, ______, ______

18. The internal angles of a square add to ______.

Monday

1. Draw the next square number in this sequence.

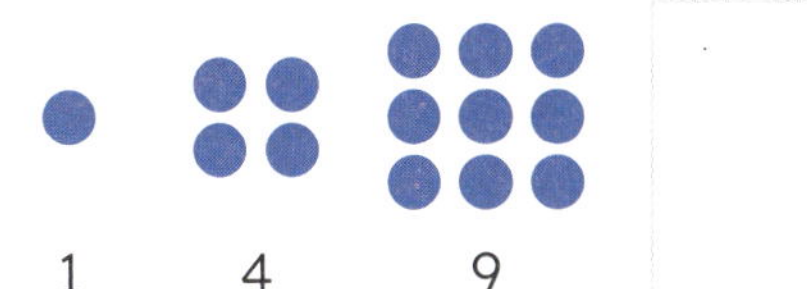

2. 9.6 = (a) $9\frac{2}{5}$ ☐ (b) $9\frac{3}{5}$ ☐ (c) $9\frac{1}{6}$ ☐

3. 17^2 = ______ × ______ = ______

4. Write $6\frac{1}{4}$ as an improper fraction. ______

5. 8% of $10.00 = ______

6.  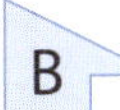  

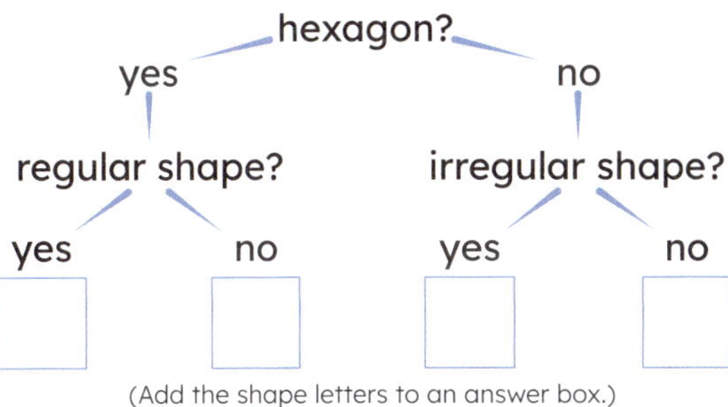

(Add the shape letters to an answer box.)

7. 75 + 85 + 65 = ______

8. $\frac{1}{1000}$m = ______mm

9. The formula for the area of a circle is πr ______.

10. If 80 × 30 = 24 × y, then y = ______.

11. $\frac{4}{5} + 2\frac{4}{5}$ = ______

12. $\frac{1}{2}$:4 = ______:8

13. Draw the line of reflectional symmetry onto the pattern.

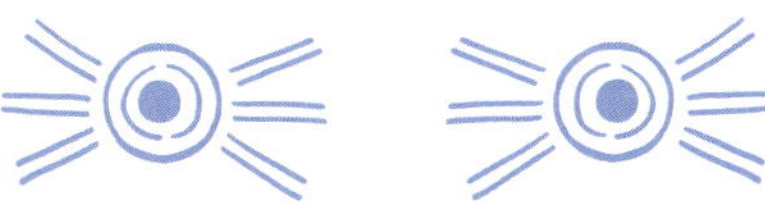

14. What is the volume of a box 40cm long, 30cm wide, and 40cm high?

______

15. What is the cost of four boxes of apples at $7.98 per kilogram if each box weighs 12kg?

______

Tuesday

Week 32

1. 5^2 = ______

2. Circle the temperature which is the anomaly.

M	Tu	W	Th	F
5°C	4°C	12°C	3°C	4°C

3. Write 3.65 million as a numeral.

______

4. If a job's rate of pay is $10 per hour, how much do you earn if you work for four hours and are paid time and a half?

______

5. Convert $7\frac{3}{4}$ to a decimal. ______.______

6. 6.05km = ______m

7. 25^2 = ______ × ______ = ______

8. If you throw a die, the chance of it landing on a six is 1 in 6. If you throw two dice, what is the chance of throwing two sixes?

______ in ______

9. Write $\frac{49}{6}$ as a mixed number. ______

10. What is the diameter of a circular driveway with a 10m radius?

______

11. A small twin-engined plane flew directly between these two towns. Using a scale of 10mm:100km, how many kilometres did the plane travel?

______

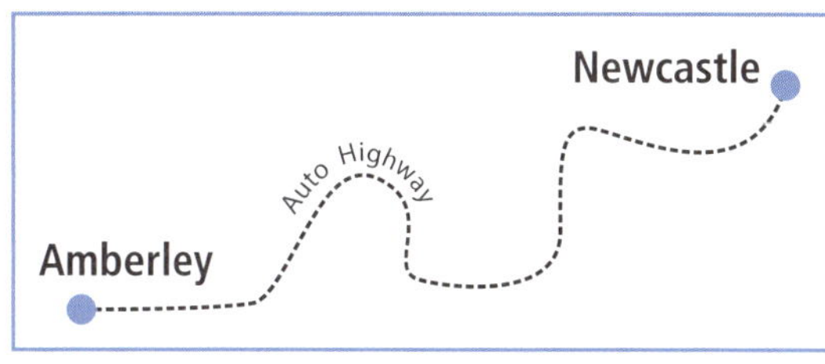

12. What is the mode of these cricket scores:

8, 10, 8, 15, 0? ______

13. 5000 ÷ (50 × 10^2) = ______

14. −9 > −2 true ☐ false ☐

15. $\frac{1}{2} \times \frac{5}{6}$ = ______

Wednesday

1. 12^2 = ________

2. The smallest angle is ________, and is opposite the smallest side, which is labelled ________.

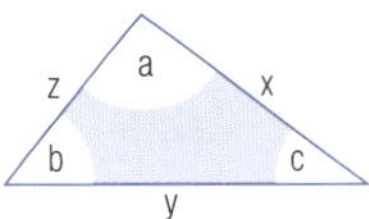

3. 7.092L = ________mL

4. If you ride 1km in four minutes, what speed are you travelling? ________

5. Simplify 12:48. ________

6. Shade the pie chart to show the favourite colours survey results.

Red	15%
Yellow	25%
Purple	30%
Blue	15%
Orange	15%

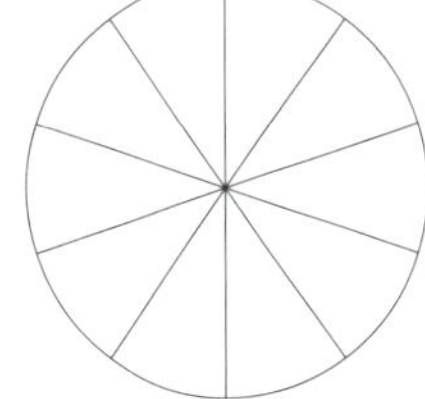

7. Write in ascending order: 1%, 0.1, $\frac{5}{10}$, 0.99.

________ ________ ________ ________

8. Which two prime numbers add up to 50?

________ and ________

9. Does this boomerang have rotational symmetry?

yes ☐ no ☐ nearly ☐

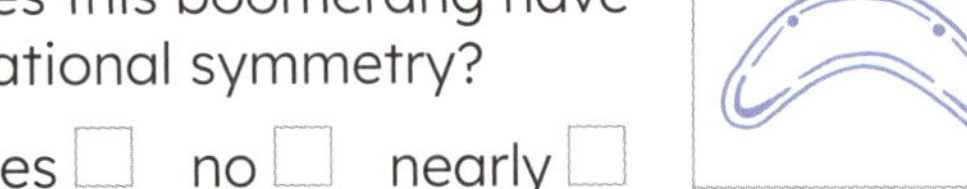

10. How many 20c coins make up $10.80? ________

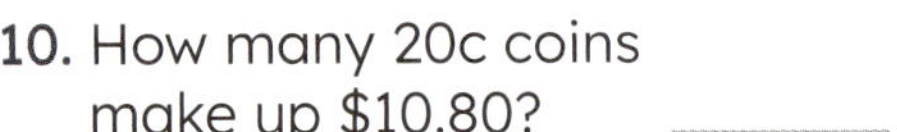

11. If $4 \times y = 280$, then y = ________.

12. Record 0.15 on the number line.

13. $100.00 – $63.55 = ________

14. Which two nets will make a cube?

________ and ________

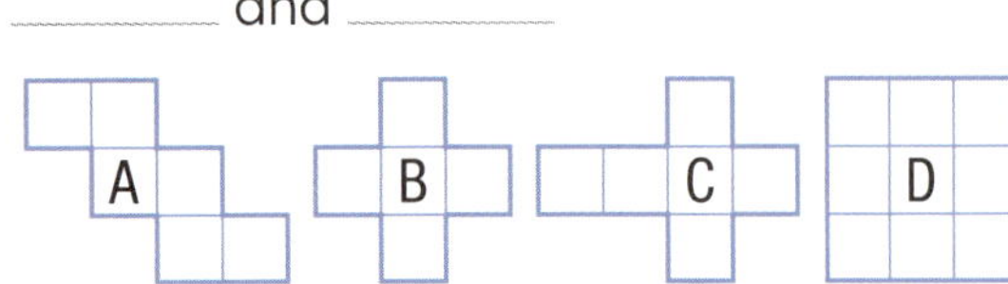

15. A bakery sold 60 loaves of bread and 32 cakes on Monday. How many items would you expect the bakery to sell in 10 days?

(a) 860 ☐ (b) 890 ☐ (c) 920 ☐

Thursday

1. 15^2 = ________

2. Follow the clues to label the bags correctly:

$A < B, C > A, B > C$

☐ ☐ ☐

3. In their backyard swamps, Alexa (a) has one-third of the amount of frogs that Kieran (k) has. Tick the expressions that show the number of frogs Alexa has.

(a) $k \times 3$ ☐ (b) $a \times 3$ ☐

(c) $\frac{k}{3}$ ☐ (d) $k + 3$ ☐

4. Draw the reflection.

457

5. 7 – 0.007 = ________

6. Write 4.64 million as a numeral.

____

7. $\frac{1}{2}$:3 = 1:________

8. 1800, 2600, 3400, ________

9. Ollie wrote the prime numbers < 20 on cards and shuffled them. What is the probability of selecting a prime number < 10?

____

10. What is the ratio of apples to pears if there are eight apples and 24 pears?

____

11. $1\frac{1}{4}$kg = 1.25kg = ________g

12. 140 – 25 = 240 – ________

13. $y°$ = ________

14. Simplify 880:1100.

____

15. 700 000 – 285 000 = ________

Problem-solving

List the square numbers from one to 20.

What do you notice about the difference between the consecutive square numbers? Display your findings using a medium of your choice.

Read the question again. **Think** about the information. Underline the important words.

Tick the strategy you will use to work out the answer:

- estimate and check ☐
- look for patterns ☐
- draw a diagram or picture ☐
- construct a table or graph ☐
- use materials ☐
- use a formula ☐
- something else. ☐

Solve it:

Reflect on the question and answer.

Check it. Circle another strategy on the list to work it out.

Show it:

Friday Review

1. 17^2 = ________
2. $y°$ = ________

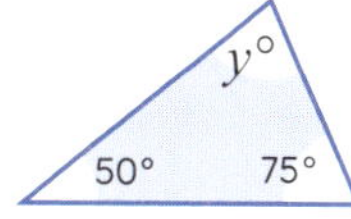

3. $\frac{2}{3} \times \frac{5}{6}$ = ________
4. $\frac{1}{2}$:6 = ________:12
5. The value of 4 in 4 378 200 is ________.
6. Rotate 450° clockwise.

A ☐

7. 6000, 12 000, 24 000, 48 000, ________
8. Where is the correct placement of 0.25?

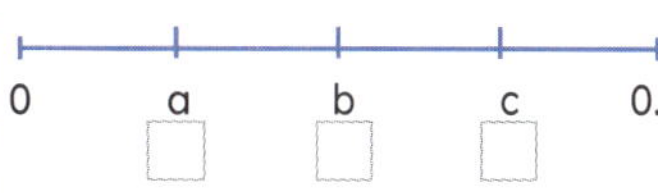

9. $3\frac{1}{3} + 3\frac{2}{3}$ = ________
10. 19^2 = ________
11. Share $200.00 equally among four people.

________ each

12. Does the witchetty grub symbol have reflectional symmetry?

yes ☐ no ☐

13. The formula for the area of a triangle is

$\frac{1}{2}$ ________ × ________.

14. If $6 \times y = 360$, then y =

________.

15. 2075mm

= ________m

16. What is the area of a 100m by 15m rectangular field?

____

17. In Thursday question 9, what is the probability of selecting a prime number > 15?

____

18. $1\frac{3}{4}$kg = ____.____kg = 1750g

Monday

1. Write the Cartesian coordinates for each point:

P = ______

Q = ______

R = ______

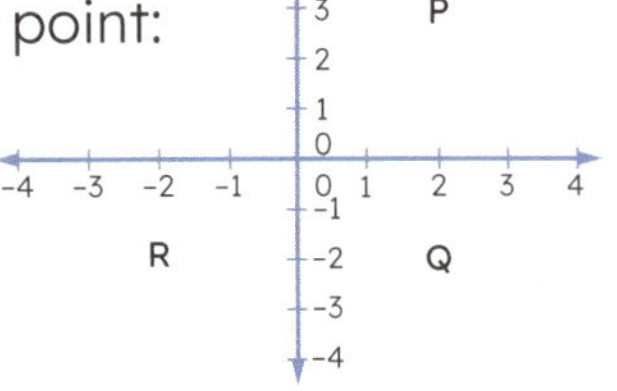

2. On a sunny day in Antarctica, the temperature at Scott's Base increased from –15°C to –5°C. By what amount did the temperature increase? ______

3. 800 – 35 = ______

4. What is the average shoe size? Huang: 7, Toto: 9, Jack: 9, Nick: 5

______

5. 45 000 + 195 000 = ______

6. Write the prime numbers found between 40 and 50. ______

7. Share $100.00 equally among 8 people. ______ each

8. A lift started on floor –2 and went up three floors. On what floor did it stop?

______

9. 16.7 × 15.8 = (a) 26.386 ☐ (b) 263.86 ☐ (c) 0.26386 ☐ (d) 2.6386 ☐

10. One floor of an apartment block has a volume of 150m^3. If all the floors have the same volume, what is the volume of a four-storey apartment building? ______m^3

11. If a + 2.3 = 3.1, then a = ______.

12. Tick the set that correctly shows which angles match.

(a) $a° = b°, c° = d°$ ☐

(b) $a° = c°, d° = b°$ ☐

(c) $b° = c°, a° = d°$ ☐

13. $\frac{3}{4}$ × 60 = ______

14. Shade the triangle with three lines of symmetry.

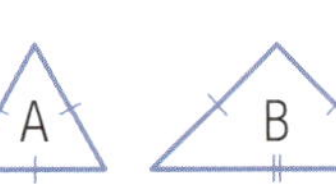
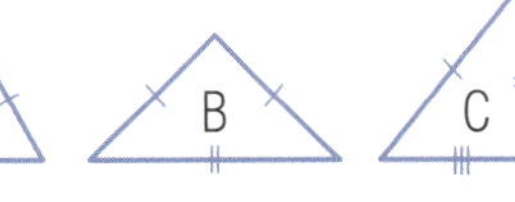

15. 7.205kg = ______g

Tuesday

1. On Monday's Cartesian plane, mark on the coordinates: (3,1), (-2,4), (-4,-3), and (3,-3).

2. An integer is a positive or negative whole number. Circle the integers below.

3 $1\frac{1}{4}$ 5 –2 1.3 7 $\frac{1}{2}$

3. What is the floor area of a 10m by 3.5m kitchen? ______

4. What is the ratio of ☐ to △? ______

5. 63 000, 71 000, 79 000, ______

6. 63 000 + 218 000 = ______

7. $\frac{1}{4}$:3 = 1:______

8. Was $12 000

now ______.

9. $\frac{3}{4} < \frac{7}{8}$

true ☐ false ☐

10. 5 × 3 × 3 = ______

11. Which is correct?

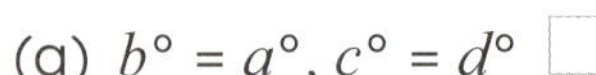
(a) $b° = a°, c° = d°$ ☐

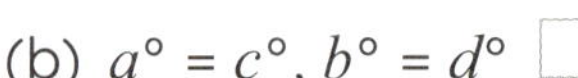
(b) $a° = c°, b° = d°$ ☐

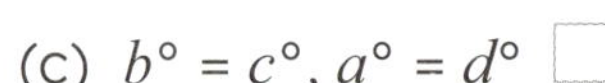
(c) $b° = c°, a° = d°$ ☐

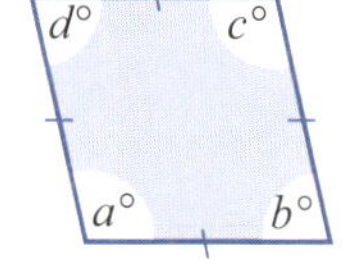

12. You are paid time and a half for working three extra hours. If the normal rate is $10 per hour, what did you earn for your overtime?

______

13. What is the probability of your name being randomly picked out of a hat containing your classmates' names?

______ in ______

14. If 80 000 = 8 × 10y, then y = ______.

15. Rotate the letter shape 540° clockwise.

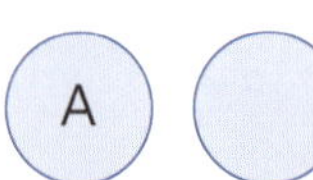

Wednesday

1. Draw the reflection of the square onto the second quadrant.

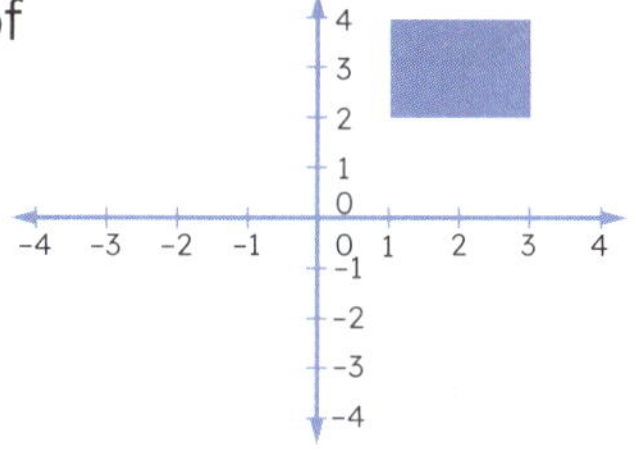

2. The coordinates for the reflected square are

(____, ____), (____, ____), (____, ____),

and (____, ____).

3. Circle the integers.

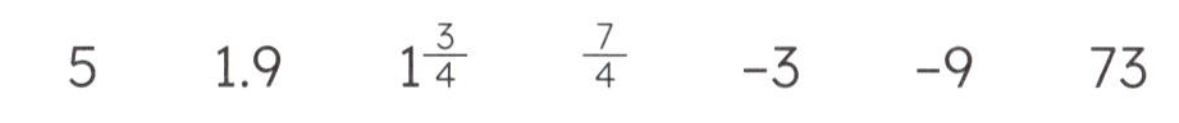

5 1.9 $1\frac{3}{4}$ $\frac{7}{4}$ −3 −9 73

4. 29.4 × 17.4 = (a) 511.56 ☐ (b) 0.51156 ☐ (c) 51.156 ☐ (d) 5115.6 ☐

5. $\frac{3}{4}$, 1, $1\frac{1}{2}$, $2\frac{1}{4}$, $3\frac{1}{4}$, ____________

6. Which is correct?

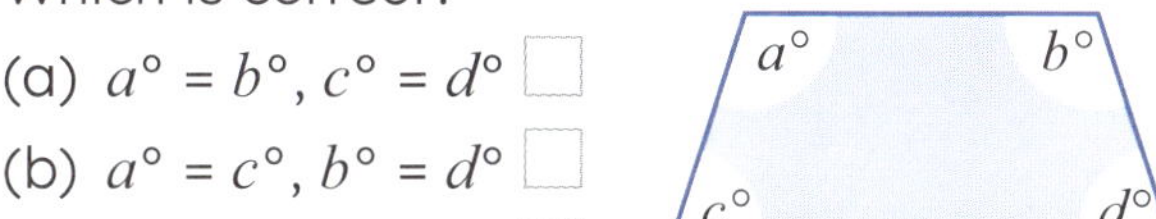

(a) $a° = b°$, $c° = d°$ ☐

(b) $a° = c°$, $b° = d°$ ☐

(c) $a° = d°$, $b° = c°$ ☐

7. What would the cost be to buy 200g of cinnamon at $2.75 per 100g?

____

8. What would you earn if you received double time for 6 hours' work at a job that normally pays $10.00 per hour?

____

9. Draw the right-side view.

10. $\frac{1}{3}$:4 = 1:____________

11. 8.005m = ____________ mm

12. What is the ratio of △ to ○? ____________

△△△○○△△△○○

13. What is the sum of the internal angles of a regular pentagon? ____________

14. If a plane departs at 0030 on Monday 12 August, what day and time must you arrive to be there 90 minutes prior?

____

15. $\frac{1}{2} \times \frac{3}{4}$ = ____________

Thursday

1. On Wednesday's Cartesian plane, draw the reflection of the square onto the fourth quadrant.

2. The coordinates for this reflected square are

(____, ____), (____, ____), (____, ____),

and (____, ____).

3. Circle the integers.

$3\frac{1}{2}$ 43 −7 2.4 92 −16

4. What is the sum of the internal angles of a hexagon?

____

5. Continue the pattern.

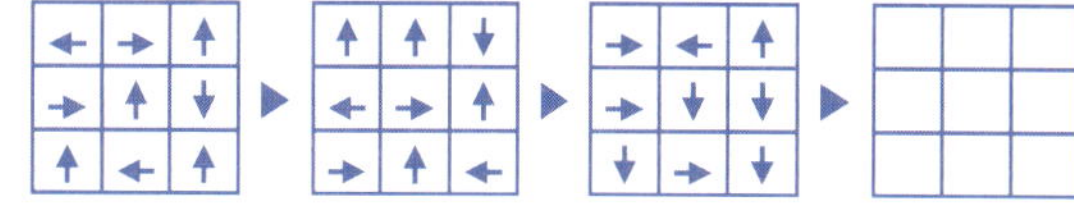

6. Write 3.05 million as a numeral.

____

7. What is the median of:

3, 3, 4, 5, 6, 8, and 9? ____________

8. 15 000, 95 000, 175 000, ____________

9. Draw the top view.

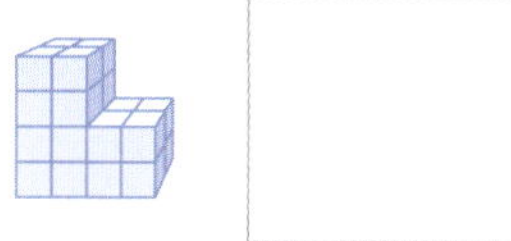

10. 14^2 = ____________

11. $\frac{1}{5}$ of an hour = ____________ minutes

12. $7\frac{1}{4} < 7.3$ true ☐ false ☐

13. Greg has in a money jar: 14 × 5c, 11 × 10c, 4 × $2, and 6 × $1. What is the total amount?

____

14. If $y \times \frac{3}{5} = \frac{3}{5}$, then y = ____________.

15. If a plane departs at 01:30 on Tuesday 11 April, what day and time must a passenger arrive to be there 90 minutes prior?

____

Problem-solving

The coordinates of one of the toes on the emu footprints is (1.5,3). Write the approximate coordinates of the five other toes on the first quadrant. Then draw the footprints reflected onto the other three quadrants.

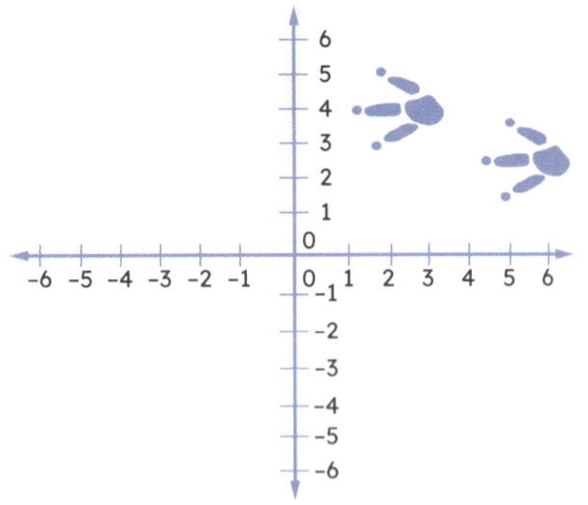

What do you notice about the coordinates of the reflections?

Read the question again. **Think** about the information. Underline the important words.

Tick the strategy you will use to work out the answer:

- estimate and check ☐
- look for patterns ☐
- draw a diagram or picture ☐
- construct a table or graph ☐
- use materials ☐
- use a formula ☐
- something else. ☐

Solve it:

Reflect on the question and answer.

Check it. Circle another strategy on the list to work it out.

Show it:

Friday Review

1. 1030 – 40 = ______

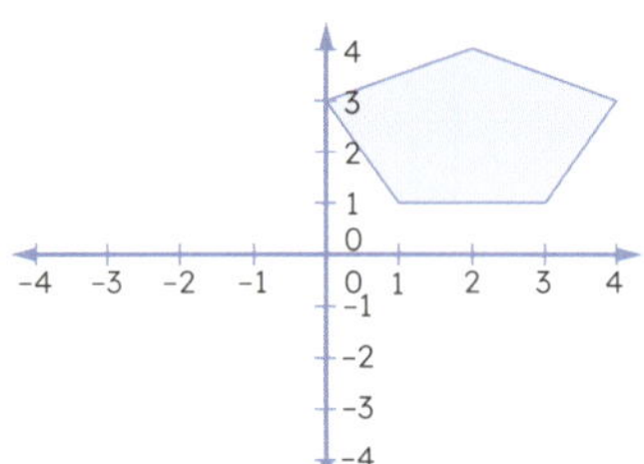

2. Draw the reflection of the pentagon onto the second quadrant.
3. The coordinates for the corners of the reflected pentagon are (___,___), (___,___), (___,___), (___,___), and (___,___).
4. Circle the integers.

   7 $3\frac{1}{2}$ –5 2.9
5. What is the ratio of □ to △? ______

6. 200 × 500 × 400 = ______
7. $\frac{1}{3}$:5 = 1:______

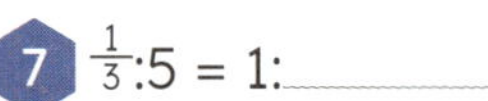

8. What is the median of 3, 3, 4, 5, 6, 10, 10, and 11? ______
9. The sum of the angles in a regular pentagon are ______.
10. 2.03m = ______cm
11. A lift started on floor –3 and went up six floors. On what floor did it stop? ______
12. The temperature in Antarctica warmed from –20°C to –10°C. What was the temperature increase? ______
13. Which angle is the same size as c? ______

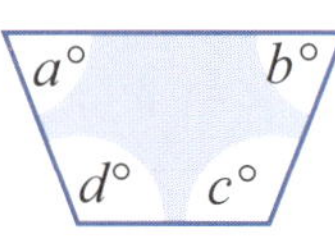

14. 7 × 8 =

    (a) 50 + 14 ☐

    (b) 20 + 26 ☐

    (c) 60 – 4 ☐

    (d) 2 + 5 × 9 ☐
15. (8000 ÷ 80) × 50 = ______
16. Draw to show a rotation of 540° clockwise.

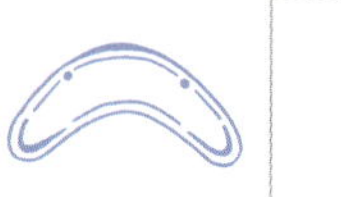

17. If you flip two coins, what is the chance of them both landing on tails? ______
18. 73 000 + 240 000 = ______

Monday

1. 9^2 = ..., $\sqrt{81}$ = ...

2. –5 + 7 = ...

3. 2 × 7 ÷ 2 + 4 = ...

4. Calculate the area of this sail.

....

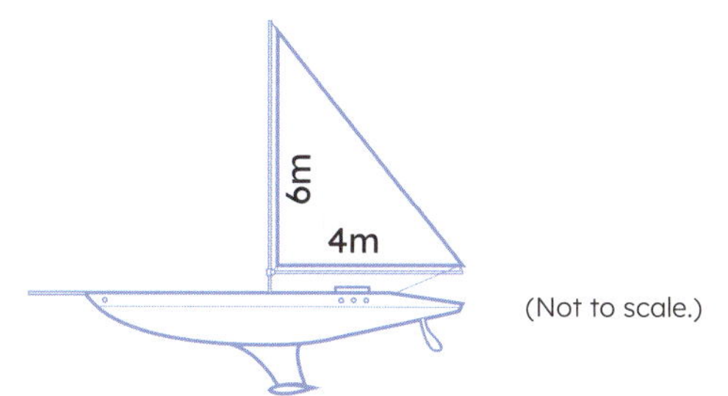

(Not to scale.)

5. 1000, 3000, 9000, 27 000, ...

6. Write the missing number.

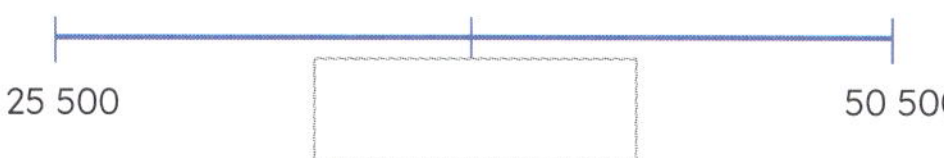

7. What is the radius of a circle which has a 50cm diameter?

....

8. If $a \div 3 = 5$, then a = ...

9. The formula for the area of a triangle is

$\frac{1}{2}$ .... × height.

10. Order from lightest to heaviest. ....

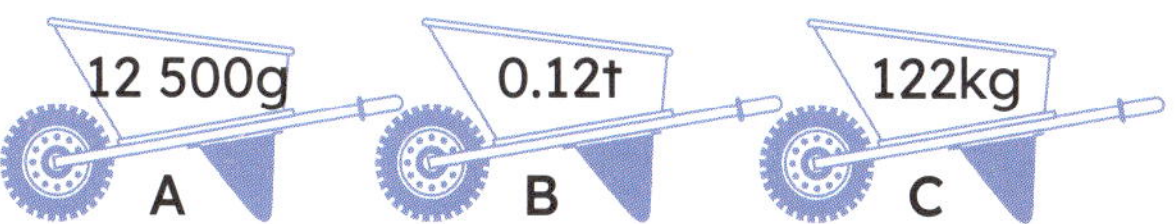

11. What is the value of 7 in 3.067?

....

12. 15 – 7 = ..., 1.5 – 0.7 = ...

13. –8 > –4 true ☐ false ☐

14. 9.38km = ...m

15. Five anglers caught a total of 29 fish. Paul caught five. The others each caught the same number of fish.

The other anglers each caught

 fish.

Tuesday

1. 6^2 = ..., $\sqrt{36}$ = ...

2. 2 – 3 = ...

3. What is the volume of soil needed to fill a rectangular trench with lengths of 3m by 2m by 1m?

....

4. 0.07 + 0.008 = ...

5. 16.8 ÷ 10 = ...

6. What marble, in which tub, has the best odds of being selected?

Tub .... Marble ....

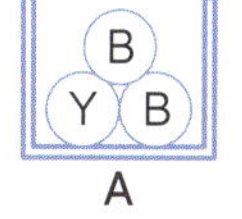

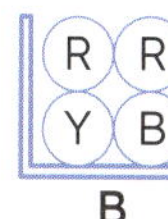

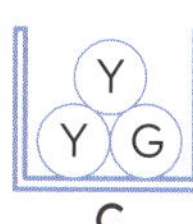

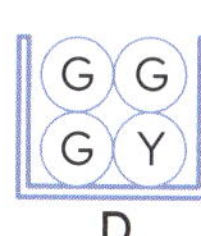

7. How many 50c coins make up \$20.50?

....

8. 30 × 10 ÷ 2 + 7 = ...

9. What is the value of the 3 in 1.073?

....

10. 0.07 × 0.3 = ...

11. $\frac{4}{5} - \frac{1}{2}$ = ...

12. Draw the reflection of the rectangle onto the second quadrant.

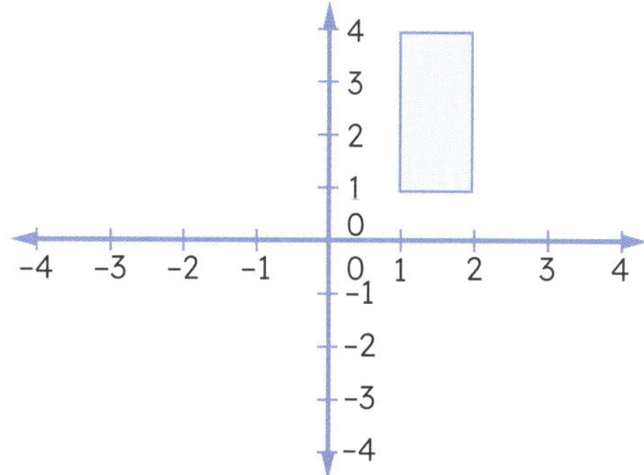

13. What is the total cost of laying a 12km-long road if it costs \$9000 per km?

....

14. If $2 \times a = 10$, then a = ...

15. 1.96, 1.97, 1.98, 1.99, ...

Week 34

Week 34

Wednesday

1. 14^2 = ______, $\sqrt{196}$ = ______

2. $\frac{1}{4}$:6 = 1:______

3. 5 – 9 = ______

4. 0.03 + 0.009 = ______

5. If 5 × a = 15, then a = ______.

6. A to B is:

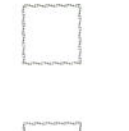

(a) an arc. ☐

(b) a tangent. ☐

(c) a chord. ☐

7. Order the following lengths from shortest to longest: 0.207km, 189m, 8000cm, 0.9m.

______ ______ ______ ______

8. Which is true?

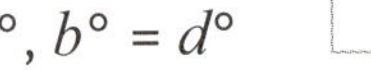

(a) $a° = c°$, $b° = d°$ ☐

(b) $a° = b°$, $c° = d°$ ☐

(c) $a° = d°$, $b° = c°$ ☐

9. 0.097, 0.098, 0.099, ______

10. Draw an irregular hexagon.

11. What is the area of this sail?

______

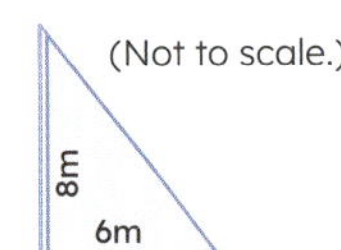

12. $\frac{2}{3} + \frac{2}{3} + \frac{2}{3}$ = ______

13. Write three capital letters that are symmetrical.

______

14. What is the ratio of

 to 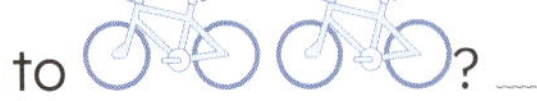? ______

15. How much would it cost to fill a 60L tank with unleaded petrol at $1.05 per litre?

______

Thursday

1. 19^2 = ______, $\sqrt{361}$ = ______

2. Circle the integers.

–9 3.4 $\frac{17}{5}$ 70 –3 9

3. What is the area of this sail? ______

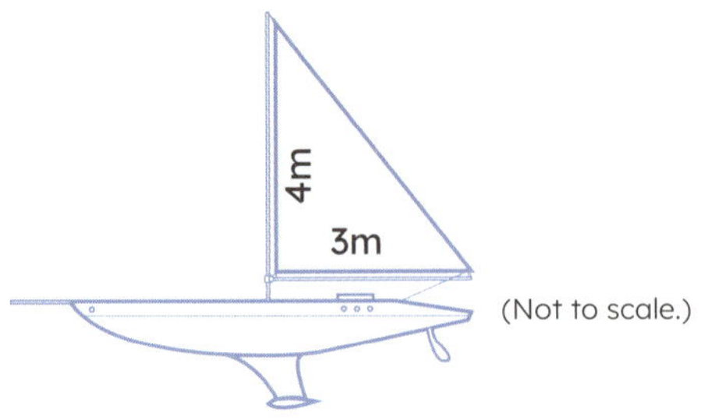

4. If a four-pack of ice cream costs $16.40, what is the price per ice cream?

______

5. Round π to the nearest whole. ______

6. The sum of angles $a°$, $b°$, and $c°$ is ______.

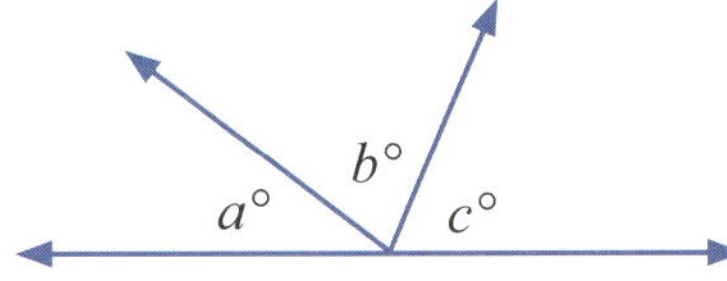

7. Simplify 35:60. ______

8. 20 × 9 = 180, 19 × 9 = ______

9. 0.05 × 0.3 = ______

10. Simplify $\frac{21}{28}$. ______

11. $\frac{1}{8}$ = 0.______

12. 6m = ______mm

13. 3.85, 3.9, 3.95, ______

14. What is the median of Hannah's hike distances: 5.5km, 3.5km, 4.1km, 2.5km, 3.5km, 2.5km, 5.4km, and 3.5km?

______

15. 14 × 30 = 2 × 7 × 30 = ______

Problem-solving

A square-shaped courtyard is paved with square slabs.

If the area of the courtyard is covered by 196 slabs, how many slab lengths form the perimeter?

Read the question again. **Think** about the information. Underline the important words.

Tick the strategy you will use to work out the answer:

- estimate and check ☐
- look for patterns ☐
- draw a diagram or picture ☐
- construct a table or graph ☐
- use materials ☐
- use a formula ☐
- something else. ☐

Solve it:

Reflect on the question and answer.

Check it. Circle another strategy on the list to work it out.

Show it:

Friday Review

Week 34

1. 18^2 = ______, $\sqrt{324}$ = ______
2. If $y \times 200 = 1000$, then y = ______.
3. What volume of concrete is needed to fill a trench 8m by 3m by 2m? ______
4. $a°$ = ______ °

   $b°$ = ______ °

   $c°$ = ______ °

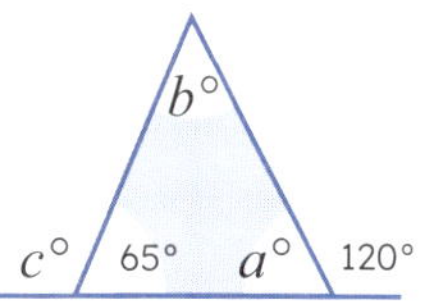

5. 0.02 + 0.006 = ______
6. What is the chance of guessing the wrong answer for question 11? ______
7. 6 – 9 = ______
8. How many 50c coins make up $31.50? ______
9. If $a^2 - 1 = 80$, then a = ______.
10. 17^2 = ______, $\sqrt{289}$ = ______
11. –7 > –3

    true ☐ false ☐

12. Show as a $\frac{3}{4}$ turn clockwise.

13. What is the range of 1.81, 1.82, 1.83, and 1.84? ______
14. If a line is drawn from a to b, what does it create?

    A ______

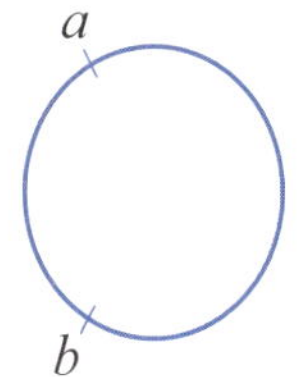

15. 200 sec. = ______ min. ______ sec.
16. 0.125, 0.25, 0.375, 0.5, ______, ______

(Not to scale.)

10m

8m

17. What is the area of this sail? ______
18. What is the cost of this sail if the material costs $10.00 per square metre? ______

N A M Sp St P

Week 35

Monday

1. $\frac{3}{4} \div \frac{1}{3} =$ ________ × ________ = ________

2. At Mirny Station in Antarctica the temperature plummeted from 1°C to –14°C. What was the drop in temperature?

____

3. $6a = 36$, $a =$ ________

4. $13^2 =$ ________, $\sqrt{169} =$ ________

5. 9900 + 700 + 800 = ________

6. Which two numbers are palindromic?

(a) 525 ☐ (b) 7297 ☐

(c) 75 257 ☐ (d) 6416 ☐

7. 10 000 – 60 = ________

8. A lift started on floor –2 and went up two floors. On what floor did it stop?

____

9. 80 + 80 000 + 900 = ________

10. Oscar caught 49 fish over 7 days. Which expression calculates the average catch per day?

(a) 49 – 7 ☐ (b) 7 ÷ 49 ☐

(c) 49 ÷ 7 ☐ (d) 49 × 7 ☐

11. How many B boxes will fit into box A?

____

(Not to scale.)

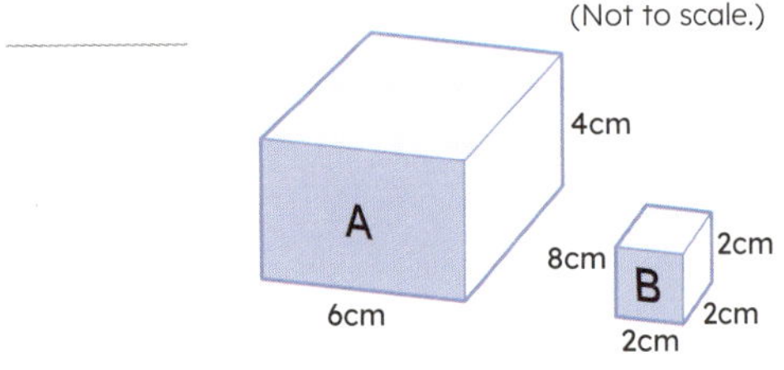

12. 4000, 400, 40, 4, ________

13. $\frac{1}{2} > 0.358$ true ☐ false ☐

14. New Wave's race car set 4 lap times at Bathurst of 2 min. 6.07 sec., 2 min. 7.39 sec., 2 min. 6.49 sec., and 2 min. 8.19 sec. What is the difference from the fastest to the slowest time?

____

15. $y° =$ ________

Tuesday

1. $\frac{1}{8} \div \frac{2}{3} =$ ________ × ________ = ________

2. $\frac{1}{3} > 0.1$ true ☐ false ☐

3. $19^2 =$ ________

4. The formula for the area of a triangle is

$\frac{1}{2}$ ________ × ________.

5. 8.03m = ________mm

6. Which line is perpendicular to D? ________

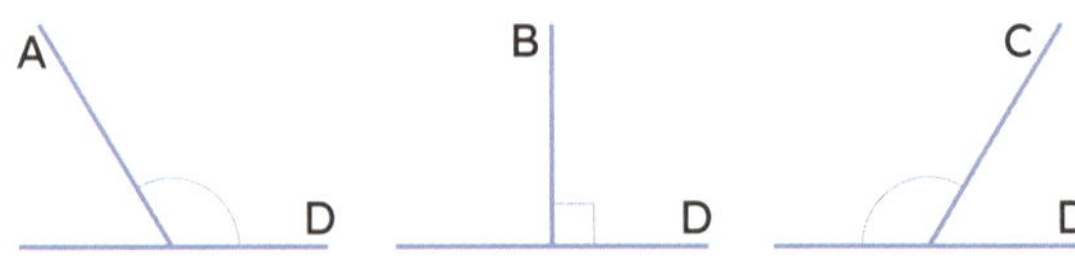

7. 0.06 + 0.007 = ________

8. One floor of an apartment block has a volume of $175m^3$. If all the floors have the same volume, what is the volume of a five-storey apartment building?

________m^3

9. 9% = 0.________

10. Which letter is at coordinate (2,–2)?________

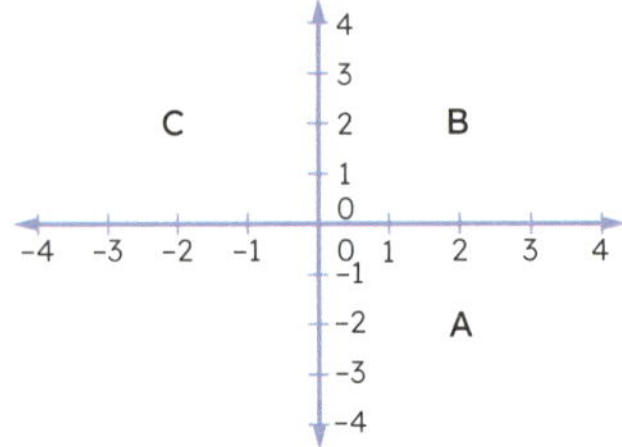

11. 900, 90, 9, 0.9, ________

12. If a plane departs on a Monday at 0500, what time should you arrive at the airport if you must be there 90 minutes earlier?

13. $y° =$ ________

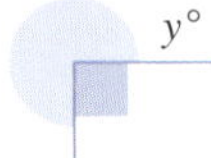

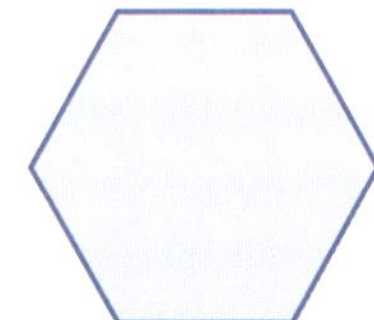

14. 10% of 20m

= 50% of ________m

15. Draw the shape's lines of symmetry.

Wednesday

1. $\frac{2}{5} \div \frac{1}{4}$ = ________ × ________ = ________

2. $\frac{1}{5} < 0.1$ true ☐ false ☐

3. 5 – 12 = ________

4. What is the total mass of these three chocolate boxes?

________kg

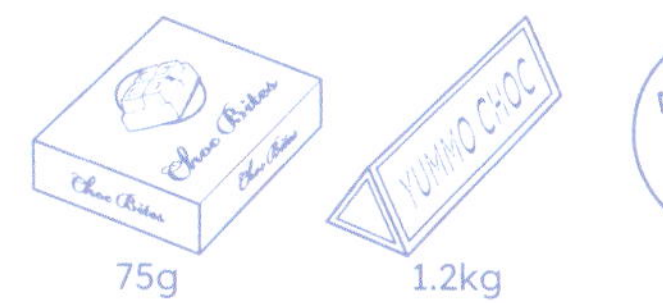

ALLCHOC SOLID CHOCOLATE BALL
200g

5. 10 005 – 30 = ________

6. 360 × 5 = 10 × ________

7. Write the angle at:

$a°$ = ________ $b°$ = ________ $c°$ = ________.

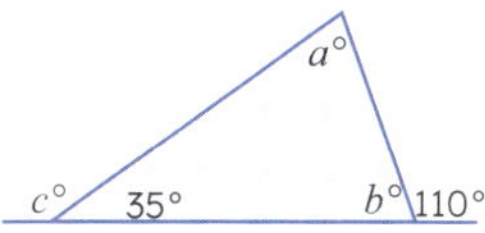

8. How many odd-numbered houses are in a street with houses numbered from 1 to 37?

____

9. 15^2 = ________, $\sqrt{225}$ = ________

10. Simplify $\frac{21}{24}$. ________

11. 400 + 45 000 + 7000 = ________

12. 10% of 30kg = 50% of ________kg

13. 4.04kg = ________g

14. A square paddock has seven fence posts along each side. How many fence posts are there altogether?

____

15. A class of 30 students each walked seven laps of an oval. Which expression matches the total number of laps the class walked?

(a) 30 + 7 ☐ (b) 30 + 7 × 2 ☐

(c) 30 × 7 ☐ (d) (30 × 7) × 7 ☐

Thursday

Week 35

1. $\frac{1}{2} \div \frac{2}{7}$ = ________ × ________ = ________

2. 8 + 6 × 4 = ________

3. A pack of three lamingtons costs $9.90. A pack of four costs $11.90. Which pack is the cheaper buy per lamington?

____

4. Double 75.5. ________

5. $\frac{1}{5}$:4 = 1:________

6. 670 × 5 = 10 × ________

7. 4×10^4 = ________

8. To travel from home to work and back again, Simon travels a total of 106km. In which two towns do they live and work?

Bunbury 149km
Busselton 86km
Nannup 139km

________ and ________

9. What is the probability of guessing the correct answer for Wednesday question 15?

____

10. 3.5 – 0.8 = ________

11. Double the size of the shaded blue shape.

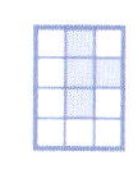

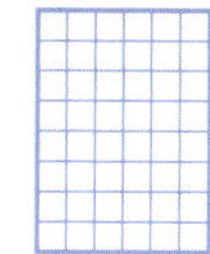

12. 7.05L = ________mL

13. How many days' difference, hours' difference, and minutes' difference are there from 2:20 pm on Tuesday to 3:45 pm on Thursday?

____

14. 4__4 + 6__ = 503
(Write the missing numbers.)

15. If you are cycling at 30km/h, how far will you ride in 20 minutes?

____

Week 35

Problem-solving

Eight equal-length sausages form a one-metre string of sausages.

How many sausages are there on a string of sausages measuring 2.75 metres?

Read the question again. **Think** about the information. Underline the important words.

Tick the strategy you will use to work out the answer:

- estimate and check ☐
- look for patterns ☐
- draw a diagram or picture ☐
- construct a table or graph ☐
- use materials ☐
- use a formula ☐
- something else. ☐

Solve it:

Reflect on the question and answer.

Check it. Circle another strategy on the list to work it out.

Show it:

Friday Review

1. $\frac{2}{5} \div \frac{3}{4} =$

   ______ × ______

   = ______

2. 2.36kg = ______ g

3. 12.2, 12.4, 12.6, 12.8, ______

4. $\frac{4}{10}$ = ______ %

5. A pack of 3 bread rolls costs $1.99. A pack of 4 costs $3.20. Which pack is cheaper per roll?

   ______

6. One floor of an apartment block has a volume of 215m^3. If all the floors have the same volume, what is the volume of a six-storey apartment building?

   ______ m^3

7. Draw the lines of symmetry.

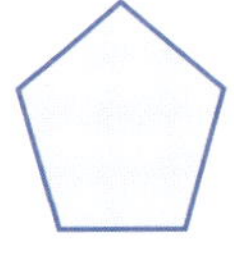

8. 17^2 = ______

9. 43 × 6 = 3 × ______

10. 12^2 = ______, $\sqrt{144}$ = ______

11. Double this shaded blue shape.

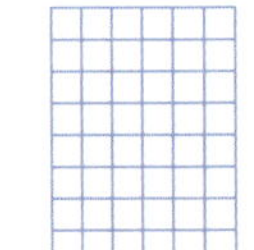

12. $\frac{1}{4} \div \frac{2}{3} =$

    ______ × ______

    = ______

13. Show a rotation of 450° clockwise.

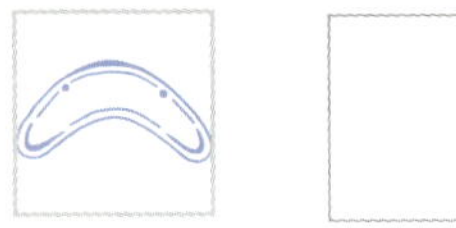

14. 74 × 8 = ______ × 4

15. What is the probability of correctly guessing tails in a one-coin flip?

    ______

16. Is A, B, C, or D at (–2,–2)?

    ______

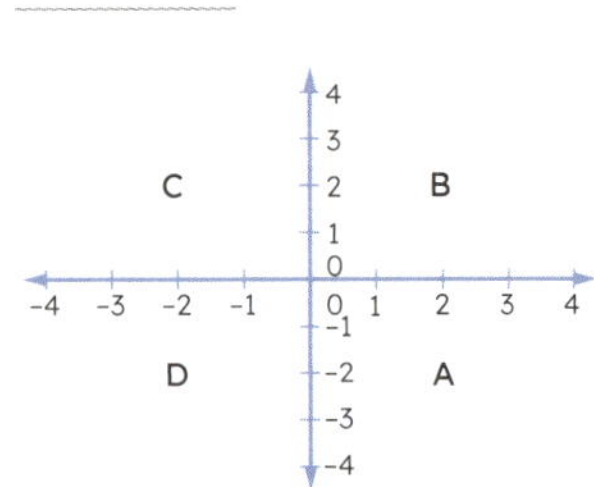

17. $a + a = 2a = 16$

    $a + a + a = 3a = 24$

    a = ______

18. Which two towns have a distance of 77km between them?

    ______

N A M Sp St P

Monday

1. Five children took a piano exam. Their results were 75%, 72%, 84%, 39%, and 87%. Which is the outlying result? ________

2. 2.3 – 0.8 = ________

3. What is the ratio of ✿ to 🌳? ________

4. A chef used a ratio of 3 parts flour to 2 parts milk for a recipe. How many cups of flour were used if the chef used 14 cups of milk? ________

5. 80 + 700 + 2000 = ________

6. $\frac{1}{4} \div \frac{5}{6}$ = ________ × ________ = ________

7. 30^2 = ________ × ________ = ________

8. Write in ascending order:
1.1, 0.05, 0.3, 2%, $\frac{4}{10}$.

________ ________ ________ ________ ________

9. Round 0.375 to the nearest tenth. ________

10. Write three capital letters that have more than one line of symmetry.

________ ________ ________

11. What is the area of this sail? ________

(Not to scale.)

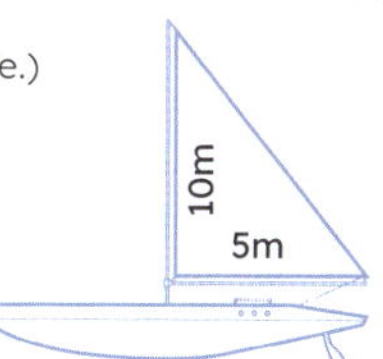

12. If sails cost $100.00 per m^2, what does the above sail cost? ________

13. 21^2 = ________

14. Draw the right-side view.

15. You have five apples and seven people. You are two apples short.

If 5 – 7 = –2, then 6 – 9 = ________.

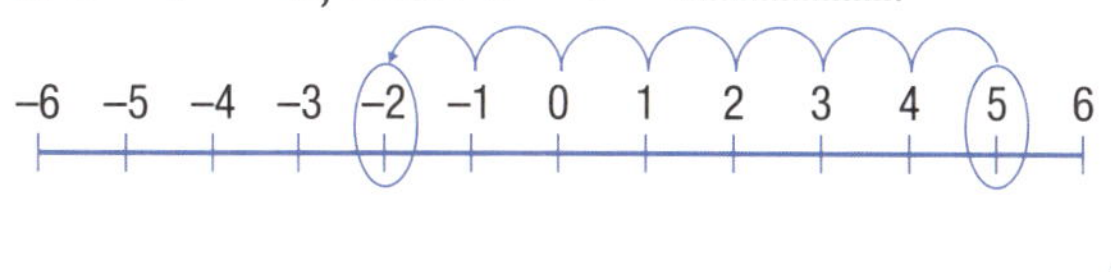

Tuesday

1. Six people were at a dinner party. Their ages were 36, 39, 44, 23, 41, and 35. Which is the outlying age? ________

2. 9 × 9 = 3 × ________

3. Circle 3 and go back five spaces. ________

4. 2400 ÷ 60 = ________

5. Write in ascending order:

10% of 2, 1.05, 4% of 4, 0.21, –5

____

6. On which number is the spinner least likely to land? ________

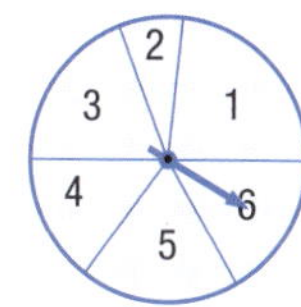

7. 2.75 × 51.3 =

(a) 1.41075 ☐ (b) 141.075 ☐

(c) 14.1075 ☐ (d) 0.141075 ☐

8. 32 ÷ 4 = 40 ÷ ________

9. 63 + 78 + 49 = ________

10. Does this shape have rotational symmetry? yes ☐ no ☐

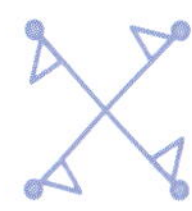

11. Write in descending order: 0.11, 0.3, 20%.

________ ________ ________

12. What is the area of this sail? ________

(Not to scale.)

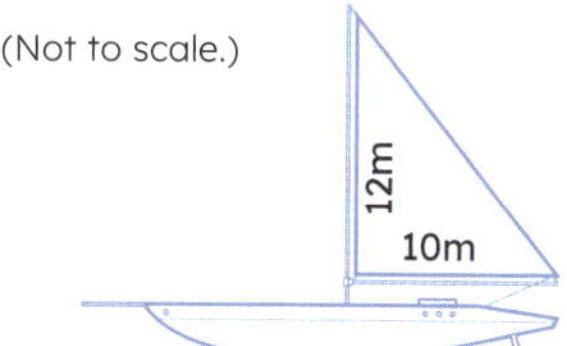

13. What is the cost of the sail if the materials cost $50.00 per m^2? ________

14. –5 < –7 ☐ true ☐ false

15. Simplify $\frac{20}{30}$. ________

Week 36

Wednesday

Week 36

1. The shoe sizes of six children were 6, 5.5, 6.5, 6, 4, and 5.5. Which is the outlying shoe size?

______

2. What is the distance between A and B?

______

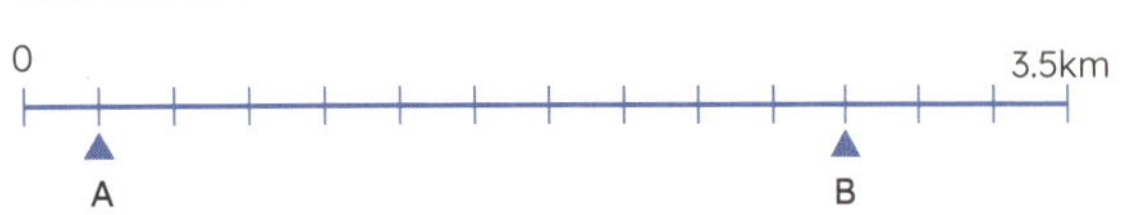

3. 9, $9\frac{1}{3}$, 10, 11, ______

4. 32 000 ÷ 40 = ______

5. 8.3 + 0.7 = ______

6. In which direction is Kiama from Jamberoo?

______

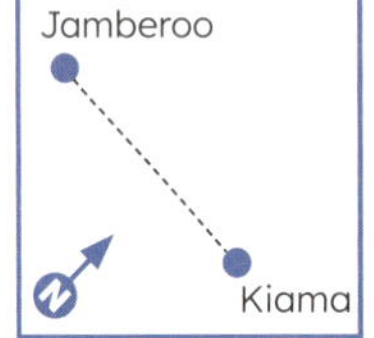

7. \$100.00 – \$36.25 = ______

8. What chance have you of going home early from school today? Record on the chance line.

0 ______ 1

9. Is 3009 divisible by 3? yes ☐ no ☐

10. The Kensington Cricket Club's batter achieved run scores of 2, 27, 100, 37, 46, 85, 51, 19, 36. What is the mean likely to be between?

(a) 80 and 90 ☐ (b) 10 and 20 ☐

(c) 0 and 10 ☐ (d) 90 and 100 ☐

(e) 40 and 50 ☐

11. $7\frac{2}{3}$ =

(a) 7.63 ☐ (b) 7.66 ☐ (c) 7.23 ☐

12. $\frac{1}{2} \times \frac{3}{4}$ = ______

13. If you are cycling at 32km/h, how far will you travel in a quarter of an hour?

______

14. Simplify 16:42. ______

15. For a party, Scott makes 6L of fruit punch. The ratio of juice to flavoured water is 5:1. How many millilitres of water is used?

______

Thursday

1. Six puppies had their weight checked by the vet. Their weights were 7kg, 6kg, 6.5kg, 8kg, 7.5kg, and 3.5kg. Which is the outlying weight?

______

2. Tiffy Pearl's famous horse scored 2019 points in an equestrian event. The next closest horse was 49 points behind. What was the 2nd horse's score?

______

3. This has rotational symmetry to the order of

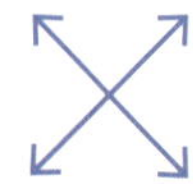

______.

4. 11 + 3 × 3 = ______

5. 132 ☐ 12 = 11

6. 909m = 0.______km

7. The chance of spinning an even number is:

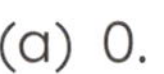

(a) 0. ☐ (b) 1. ☐

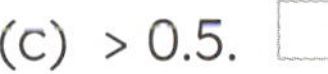

(c) > 0.5. ☐ (d) < 0.5. ☐

8. 20^2 = ______, $\sqrt{400}$ = ______

9. 12 × 7 = ______

10. 2024 – 21 = ______

11. If a bus leaves its depot at 6:04 am and has 12 stops at 4 minute intervals, at what time will it arrive at the 12th stop?

______

12. 96 × 6 = 3 × ______

13. If $\frac{500}{e}$ = 100, then e = ______.

14. Order the boxes from lightest to heaviest.

______ ______ ______

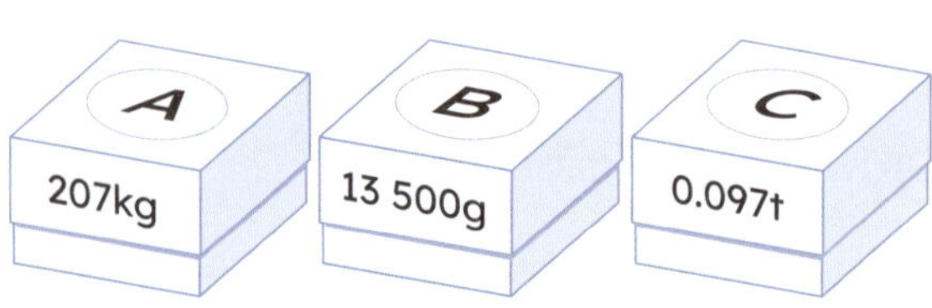

15. Simplify $3\frac{16}{20}$. ______

Problem-solving

Survey your class about an aspect of their lives that can be counted. For example, how many pets or siblings they have.

Display your findings and identify any outliers.

Read the question again. **Think** about the information. Underline the important words.

Tick the strategy you will use to work out the answer:

- estimate and check ☐
- look for patterns ☐
- draw a diagram or picture ☐
- construct a table or graph ☐
- use materials ☐
- use a formula ☐
- something else. ☐

Solve it:

Reflect on the question and answer.

Check it. Circle another strategy on the list to work it out.

Show it:

Friday Review

1. The results in a maths test were 65%, 90%, 71%, 68%, 27%, and 75%. Which is the outlying result?

   ____

2. 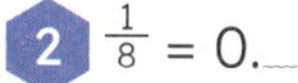 $\frac{1}{8}$ = 0.________

3. 36 ÷ 4 × 3 = ________

4. The chance of spinning an odd number is:

   (a) > 0.5. ☐

   (b) < 0.5. ☐

   (c) 1. ☐

   (d) 1 in 4. ☐

   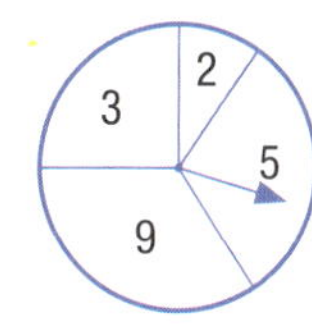

5. 2070mm = ________m

6. If $\frac{600}{e}$ = 100, then e = ________.

7. 5, $5\frac{1}{4}$, $5\frac{3}{4}$, $6\frac{1}{2}$,

   ____

8. 98 × 8 = ________ × 4

9. 9 × 10^4 = ________

10. Share $500 equally among 20 people.

    ________ each

11. Round 0.477 to the nearest tenth.

    ____

12. Does this shape have rotational symmetry?

    ____

13. Write in descending order:

    0.32, 15%, 2.1, $\frac{1}{5}$ of 4, 72 × 0.01.

    ____

    ____

14. 72  6 = 12

15. Five students did a sponsored hike. Their distances were 5km, 5.5km, 4.8km, 3.2km, and 5.7km. Which is the outlying distance?

    ____

16.  $8\frac{4}{5}$ = ________.________

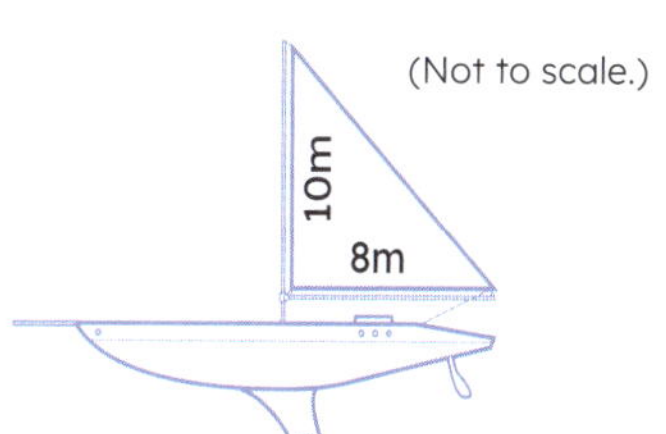

(Not to scale.)

17. What is the area of this sail?

    ____

18. What is the cost of the above sail if a sail maker charges $20 per square metre?

    ____

Maths Facts

Number

Place value

9741.25
9000.00
700.00
40.00
1.00
0.20
0.05

thousands	hundreds	tens	ones	•	tenths	hundredths
9	7	4	1	.	2	5

Fractions

Numerator

The number above the line, indicating how many parts of the whole are in consideration.

Denominator

The number below the line, indicating how many parts the whole number is divided into.

Equivalent fractions

<table>
<tr><td colspan="360">one whole</td></tr>
<tr><td colspan="180">$\frac{1}{2}$</td><td colspan="180">$\frac{1}{2}$</td></tr>
<tr><td colspan="90">$\frac{1}{4}$</td><td colspan="90">$\frac{1}{4}$</td><td colspan="90">$\frac{1}{4}$</td><td colspan="90">$\frac{1}{4}$</td></tr>
<tr><td colspan="45">$\frac{1}{8}$</td><td colspan="45">$\frac{1}{8}$</td><td colspan="45">$\frac{1}{8}$</td><td colspan="45">$\frac{1}{8}$</td><td colspan="45">$\frac{1}{8}$</td><td colspan="45">$\frac{1}{8}$</td><td colspan="45">$\frac{1}{8}$</td><td colspan="45">$\frac{1}{8}$</td></tr>
<tr><td colspan="120">$\frac{1}{3}$</td><td colspan="120">$\frac{1}{3}$</td><td colspan="120">$\frac{1}{3}$</td></tr>
<tr><td colspan="60">$\frac{1}{6}$</td><td colspan="60">$\frac{1}{6}$</td><td colspan="60">$\frac{1}{6}$</td><td colspan="60">$\frac{1}{6}$</td><td colspan="60">$\frac{1}{6}$</td><td colspan="60">$\frac{1}{6}$</td></tr>
<tr><td colspan="40">$\frac{1}{9}$</td><td colspan="40">$\frac{1}{9}$</td><td colspan="40">$\frac{1}{9}$</td><td colspan="40">$\frac{1}{9}$</td><td colspan="40">$\frac{1}{9}$</td><td colspan="40">$\frac{1}{9}$</td><td colspan="40">$\frac{1}{9}$</td><td colspan="40">$\frac{1}{9}$</td><td colspan="40">$\frac{1}{9}$</td></tr>
<tr><td colspan="30">$\frac{1}{12}$</td><td colspan="30">$\frac{1}{12}$</td><td colspan="30">$\frac{1}{12}$</td><td colspan="30">$\frac{1}{12}$</td><td colspan="30">$\frac{1}{12}$</td><td colspan="30">$\frac{1}{12}$</td><td colspan="30">$\frac{1}{12}$</td><td colspan="30">$\frac{1}{12}$</td><td colspan="30">$\frac{1}{12}$</td><td colspan="30">$\frac{1}{12}$</td><td colspan="30">$\frac{1}{12}$</td><td colspan="30">$\frac{1}{12}$</td></tr>
<tr><td colspan="72">$\frac{1}{5}$</td><td colspan="72">$\frac{1}{5}$</td><td colspan="72">$\frac{1}{5}$</td><td colspan="72">$\frac{1}{5}$</td><td colspan="72">$\frac{1}{5}$</td></tr>
<tr><td colspan="36">$\frac{1}{10}$</td><td colspan="36">$\frac{1}{10}$</td><td colspan="36">$\frac{1}{10}$</td><td colspan="36">$\frac{1}{10}$</td><td colspan="36">$\frac{1}{10}$</td><td colspan="36">$\frac{1}{10}$</td><td colspan="36">$\frac{1}{10}$</td><td colspan="36">$\frac{1}{10}$</td><td colspan="36">$\frac{1}{10}$</td><td colspan="36">$\frac{1}{10}$</td></tr>
</table>

Symbols

$\sqrt{}$ means square root
$+$ means add
$-$ means subtract
$\times$ means multiply
$\div$ means divide
$=$ means equal to
$<$ means less than
$>$ means greater than

Prime and composite numbers

Prime numbers up to 20

2, 3, 5, 7, 11, 13, 17, and 19

Composite numbers up to 20

4, 6, 8, 9, 10, 12, 14, 15, 16, 18, and 20

Maths Facts

Number

Multiplication and division facts

Multiplication and division facts are linked together. If you know one multiplication fact, you also have the ability to know three other related facts.

For example:
If you know the fact 4 × 5 = 20,
you will also know the facts 5 × 4 = 20, 20 ÷ 4 = 5, and 20 ÷ 5 = 4.

2 × table	Other facts I know		
1 × 2 = 2	2 × 1 = 2	2 ÷ 2 = 1	2 ÷ 1 = 2
2 × 2 = 4		4 ÷ 2 = 2	
3 × 2 = 6	2 × 3 = 6	6 ÷ 2 = 3	6 ÷ 3 = 2
4 × 2 = 8	2 × 4 = 8	8 ÷ 2 = 4	8 ÷ 4 = 2
5 × 2 = 10	2 × 5 = 10	10 ÷ 2 = 5	10 ÷ 5 = 2
6 × 2 = 12	2 × 6 = 12	12 ÷ 2 = 6	12 ÷ 6 = 2
7 × 2 = 14	2 × 7 = 14	14 ÷ 2 = 7	14 ÷ 7 = 2
8 × 2 = 16	2 × 8 = 16	16 ÷ 2 = 8	16 ÷ 8 = 2
9 × 2 = 18	2 × 9 = 18	18 ÷ 2 = 9	18 ÷ 9 = 2
10 × 2 = 20	2 × 10 = 20	20 ÷ 2 = 10	20 ÷ 10 = 2

3 × table	Other facts I know		
1 × 3 = 3	3 × 1 = 3	3 ÷ 3 = 1	3 ÷ 1 = 3
2 × 3 = 6	3 × 2 = 6	6 ÷ 3 = 2	6 ÷ 2 = 3
3 × 3 = 9		9 ÷ 3 = 3	
4 × 3 = 12	3 × 4 = 12	12 ÷ 3 = 4	12 ÷ 4 = 3
5 × 3 = 15	3 × 5 = 15	15 ÷ 3 = 5	15 ÷ 5 = 3
6 × 3 = 18	3 × 6 = 18	18 ÷ 3 = 6	18 ÷ 6 = 3
7 × 3 = 21	3 × 7 = 21	21 ÷ 3 = 7	21 ÷ 7 = 3
8 × 3 = 24	3 × 8 = 24	24 ÷ 3 = 8	24 ÷ 8 = 3
9 × 3 = 27	3 × 9 = 27	27 ÷ 3 = 9	27 ÷ 9 = 3
10 × 3 = 30	3 × 10 = 30	30 ÷ 3 = 10	30 ÷ 10 = 3

4 × table	Other facts I know		
1 × 4 = 4	4 × 1 = 4	4 ÷ 4 = 1	4 ÷ 1 = 4
2 × 4 = 8	4 × 2 = 8	8 ÷ 4 = 2	8 ÷ 2 = 4
3 × 4 = 12	4 × 3 = 12	12 ÷ 4 = 3	12 ÷ 3 = 4
4 × 4 = 16		16 ÷ 4 = 4	
5 × 4 = 20	4 × 5 = 20	20 ÷ 4 = 5	20 ÷ 5 = 4
6 × 4 = 24	4 × 6 = 24	24 ÷ 4 = 6	24 ÷ 6 = 4
7 × 4 = 28	4 × 7 = 28	28 ÷ 4 = 7	28 ÷ 7 = 4
8 × 4 = 32	4 × 8 = 32	32 ÷ 4 = 8	32 ÷ 8 = 4
9 × 4 = 36	4 × 9 = 36	36 ÷ 4 = 9	36 ÷ 9 = 4
10 × 4 = 40	4 × 10 = 40	40 ÷ 4 = 10	40 ÷ 10 = 4

5 × table	Other facts I know		
1 × 5 = 5	5 × 1 = 5	5 ÷ 5 = 1	5 ÷ 1 = 5
2 × 5 = 10	5 × 2 = 10	10 ÷ 5 = 2	10 ÷ 2 = 5
3 × 5 = 15	5 × 3 = 15	15 ÷ 5 = 3	15 ÷ 3 = 5
4 × 5 = 20	5 × 4 = 20	20 ÷ 5 = 4	20 ÷ 4 = 5
5 × 5 = 25		25 ÷ 5 = 5	
6 × 5 = 30	5 × 6 = 30	30 ÷ 5 = 6	30 ÷ 6 = 5
7 × 5 = 35	5 × 7 = 35	35 ÷ 5 = 7	35 ÷ 7 = 5
8 × 5 = 40	5 × 8 = 40	40 ÷ 5 = 8	40 ÷ 8 = 5
9 × 5 = 45	5 × 9 = 45	45 ÷ 5 = 9	45 ÷ 9 = 5
10 × 5 = 50	5 × 10 = 50	50 ÷ 5 = 10	50 ÷ 10 = 5

6 × table	Other facts I know		
1 × 6 = 6	6 × 1 = 6	6 ÷ 6 = 1	6 ÷ 1 = 6
2 × 6 = 12	6 × 2 = 12	12 ÷ 6 = 2	12 ÷ 2 = 6
3 × 6 = 18	6 × 3 = 18	18 ÷ 6 = 3	18 ÷ 3 = 6
4 × 6 = 24	6 × 4 = 24	24 ÷ 6 = 4	24 ÷ 4 = 6
5 × 6 = 30	6 × 5 = 30	30 ÷ 6 = 5	30 ÷ 5 = 6
6 × 6 = 36		36 ÷ 6 = 6	
7 × 6 = 42	6 × 7 = 42	42 ÷ 6 = 7	42 ÷ 7 = 6
8 × 6 = 48	6 × 8 = 48	48 ÷ 6 = 8	48 ÷ 8 = 6
9 × 6 = 54	6 × 9 = 54	54 ÷ 6 = 9	54 ÷ 9 = 6
10 × 6 = 60	6 × 10 = 60	60 ÷ 6 = 10	60 ÷ 10 = 6

7 × table	Other facts I know		
1 × 7 = 7	7 × 1 = 7	7 ÷ 7 = 1	7 ÷ 1 = 7
2 × 7 = 14	7 × 2 = 14	14 ÷ 7 = 2	14 ÷ 2 = 7
3 × 7 = 21	7 × 3 = 21	21 ÷ 7 = 3	21 ÷ 3 = 7
4 × 7 = 28	7 × 4 = 28	28 ÷ 7 = 4	28 ÷ 4 = 7
5 × 7 = 35	7 × 5 = 35	35 ÷ 7 = 5	35 ÷ 5 = 7
6 × 7 = 42	7 × 6 = 42	42 ÷ 7 = 6	42 ÷ 6 = 7
7 × 7 = 49		49 ÷ 7 = 7	
8 × 7 = 56	7 × 8 = 56	56 ÷ 7 = 8	56 ÷ 8 = 7
9 × 7 = 63	7 × 9 = 63	63 ÷ 7 = 9	63 ÷ 9 = 7
10 × 7 = 70	7 × 10 = 70	70 ÷ 7 = 10	70 ÷ 10 = 7

Maths Facts

Number

8 × table	Other facts I know		
1 × 8 = 8	8 × 1 = 8	8 ÷ 8 = 1	8 ÷ 1 = 8
2 × 8 = 16	8 × 2 = 16	16 ÷ 8 = 2	16 ÷ 2 = 8
3 × 8 = 24	8 × 3 = 24	24 ÷ 8 = 3	24 ÷ 3 = 8
4 × 8 = 32	8 × 4 = 32	32 ÷ 8 = 4	32 ÷ 4 = 8
5 × 8 = 40	8 × 5 = 40	40 ÷ 8 = 5	40 ÷ 5 = 8
6 × 8 = 48	8 × 6 = 48	48 ÷ 8 = 6	48 ÷ 6 = 8
7 × 8 = 56	8 × 7 = 56	56 ÷ 8 = 7	56 ÷ 7 = 8
8 × 8 = 64		64 ÷ 8 = 8	
9 × 8 = 72	8 × 9 = 72	72 ÷ 8 = 9	72 ÷ 9 = 8
10 × 8 = 80	8 × 10 = 80	80 ÷ 8 = 10	80 ÷ 10 = 8

9 × table	Other facts I know		
1 × 9 = 9	9 × 1 = 9	9 ÷ 9 = 1	9 ÷ 1 = 9
2 × 9 = 18	9 × 2 = 18	18 ÷ 9 = 2	18 ÷ 2 = 9
3 × 9 = 27	9 × 3 = 27	27 ÷ 9 = 3	27 ÷ 3 = 9
4 × 9 = 36	9 × 4 = 36	36 ÷ 9 = 4	36 ÷ 4 = 9
5 × 9 = 45	9 × 5 = 45	45 ÷ 9 = 5	45 ÷ 5 = 9
6 × 9 = 54	9 × 6 = 54	54 ÷ 9 = 6	54 ÷ 6 = 9
7 × 9 = 63	9 × 7 = 63	63 ÷ 9 = 7	63 ÷ 7 = 9
8 × 9 = 72	9 × 8 = 72	72 ÷ 9 = 8	72 ÷ 8 = 9
9 × 9 = 81		81 ÷ 9 = 9	
10 × 9 = 90	9 × 10 = 90	90 ÷ 9 = 10	90 ÷ 10 = 9

10 × table	Other facts I know		
1 × 10 = 10	10 × 1 = 10	10 ÷ 10 = 1	10 ÷ 1 = 10
2 × 10 = 20	10 × 2 = 20	20 ÷ 10 = 2	20 ÷ 2 = 10
3 × 10 = 30	10 × 3 = 30	30 ÷ 10 = 3	30 ÷ 3 = 10
4 × 10 = 40	10 × 4 = 40	40 ÷ 10 = 4	40 ÷ 4 = 10
5 × 10 = 50	10 × 5 = 50	50 ÷ 10 = 5	50 ÷ 5 = 10
6 × 10 = 60	10 × 6 = 60	60 ÷ 10 = 6	60 ÷ 6 = 10
7 × 10 = 70	10 × 7 = 70	70 ÷ 10 = 7	70 ÷ 7 = 10
8 × 10 = 80	10 × 8 = 80	80 ÷ 10 = 8	80 ÷ 8 = 10
9 × 10 = 90	10 × 9 = 90	90 ÷ 10 = 9	90 ÷ 9 = 10
10 × 10 = 100		100 ÷ 10 = 10	

Cross Strand Number and Measurement

What coins make one dollar?

1 DOLLAR

20 × 5

10 × 10

5 × 20

2 × 50

Algebra

Order of operations – BODMAS

There are rules to follow with complicated calculations that have several operations or brackets; the parts need to be done in the following order:

Brackets
Orders (powers and square roots)
Division *
Multiplication *
Addition *
Subtraction *

*Left to right, whichever of the two operations comes first.

Maths Facts

Measurement

Length

Unit	Abbreviation
millimetre	mm
centimetre	cm
metre	m
kilometre	km

10mm = 1cm

100cm = 1m

1000m = 1km

The circle

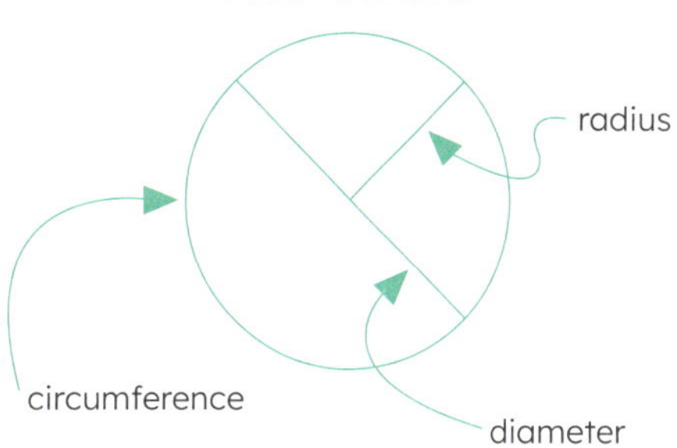

Radius: a straight line that joins the centre of a circle to a point on the circle's outer edge.

Circumference: the distance of a circle's boundary (its perimeter).

Diameter: a straight line that passes through the centre of a circle, dividing the circle into halves. The diameter is twice the length of the radius (d = 2 × r).

Angles

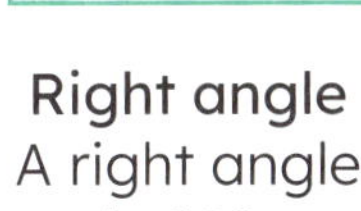

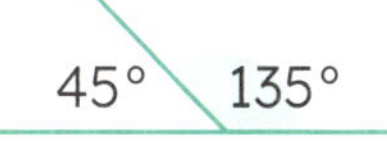

45° 135°

35° 55°

Acute angle An acute angle is less than 90°.

Right angle A right angle is 90°.

Obtuse angle An obtuse angle is between 90° and 180°.

If two angles add to 180°, they are known as supplementary.

If two angles add to 90°, they are known as complementary.

Area

The area of a rectangle can be found by applying the formula:

area = length × width

Examples

3cm, 2cm

Area = L × W
Area = 3cm × 2cm
Area = 6cm^2

6m, 1m

Area = L × W
Area = 6m × 1m
Area = 6m^2

The area of irregular shapes can be found by breaking the area into appropriate shapes; for example, rectangles.

Area = (L × W) + (L × W)
Area = (3cm × 2cm) + (6cm × 2cm)
Area = 6cm^2 + 12cm^2
Area = 18cm^2

1 hectare = 10 000m^2

Maths Facts

Measurement

Mass

Unit	Abbreviation
gram	g
kilogram	kg
tonne	t

1000g = 1kg | 1000kg = 1t

Capacity

Unit	Abbreviation
millilitre	mL
litre	L

1000mL = 1L

Money

Unit	Symbol
cent	c
dollar	$

100c = $1.00

Time

24-hour clock

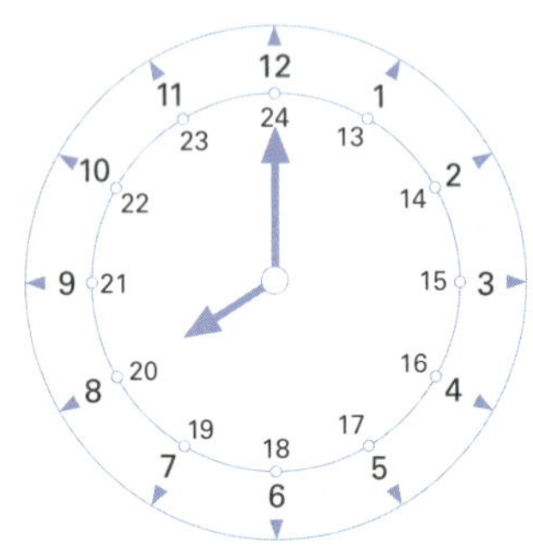

am
The time between midnight and midday is am time.

pm
The time between midday and midnight is pm time.

Examples:

8 am = 0800 hours
8 pm = 2000 hours
2:30 am = 0230 hours
2:30 pm = 1430 hours
12:01 am = 0001 hours
11:59 pm = 2359 hours

Time zones

Australia is divided into three time zones.

- **Eastern Standard Time** (AEST) includes Queensland, New South Wales (excluding Broken Hill), Victoria, Tasmania, and the Australian Capital Territory. AEST is equal to Coordinated Universal Time plus ten hours (UTC+10:00).
- **Central Standard Time** (ACST) includes South Australia, Broken Hill, and the Northern Territory. ACST is equal to Coordinated Universal Time plus nine and a half hours (UTC+09:30).
- **Western Standard Time** (AWST) includes Western Australia. AWST is equal to Coordinated Universal Time plus eight hours (UTC+08:00).

Daylight Saving Time (ADST) involves advancing clocks by one hour in warmer months. This is observed in New South Wales (excluding Broken Hill), Victoria, South Australia, Tasmania, and the Australian Capital Territory.

- **Eastern Daylight Time** (AEDT) is observed in New South Wales, Victoria, Tasmania, and the Australian Capital Territory. The clocks advance to UTC+11:00.
- Daylight saving is not observed in Queensland, the Northern Territory, or Western Australia.
- **Central Daylight Time** (ACDT) is observed in South Australia and Broken Hill. The clocks advance to UTC+10:30.

Cross Strand Measurement and Space

Lines

Horizontal
A line parallel to the horizon.

Vertical
A line which is at right angles to a horizontal line.

Parallel
Lines that are always the same distance apart and have no common points.

Perpendicular
A line at a right angle to another line.

Oblique
A slanted line that is not horizontal or vertical.

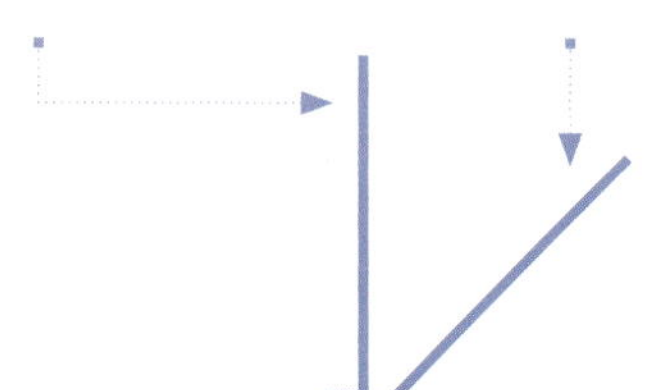

Maths Facts

Space

2D shapes – triangles

A triangle is a shape with 3 sides and 3 angles. The total of the angles adds up to 180°.

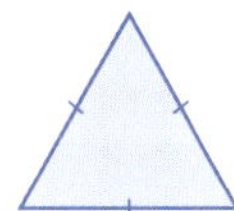

equilateral triangle
3 sides the same length.
3 angles the same size.

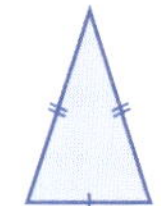

isosceles triangle
2 sides the same length.
2 angles the same size.

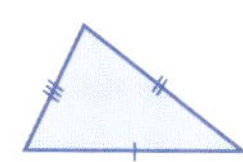

scalene triangle
0 sides the same length.
0 angles the same size.

2D shapes – quadrilaterals

A quadrilateral is a shape with 4 sides and 4 angles. The total of the angles adds up to 360°.

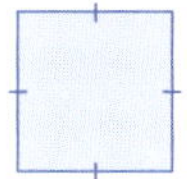

square
4 sides the same length.
4 right angles.

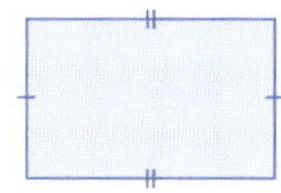

rectangle
2 pairs of sides the same length.
4 right angles.

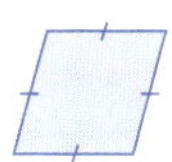

rhombus
4 sides the same length.
Opposite angles are equal.

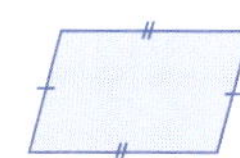

parallelogram
2 pairs of sides the same length.
Opposite angles are equal.

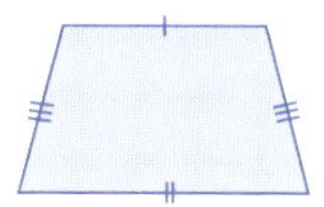

trapezium
1 pair of sides the same length.
2 pairs of angles the same size.

Other 2D shapes

Corner: Where two straight lines meet.

circle
1 curved side
0 corners

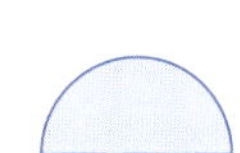

semicircle
1 curved side
1 straight side
0 corners

ellipse
1 curved side
0 corners

pentagon
5 sides
5 corners

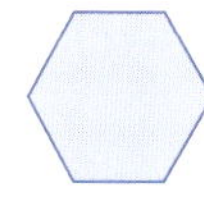

hexagon
6 sides
6 corners

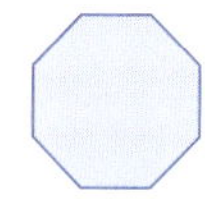

octagon
8 sides
8 corners

3D objects

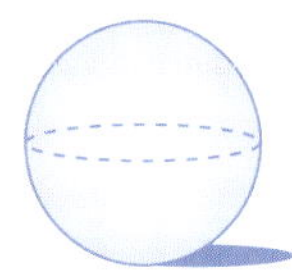

sphere
0 faces 1 surface
0 edges 0 vertices

cone
1 face 1 surface
1 curved edge
1 vertex

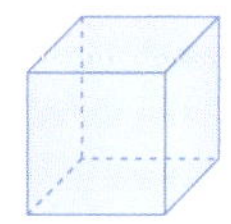

cube
6 faces 12 edges
8 vertices

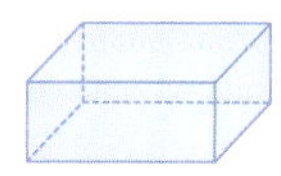

rectangular prism
6 faces 12 edges
8 vertices

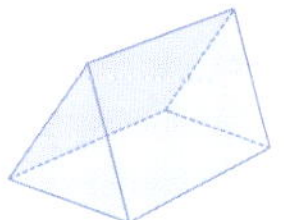

triangular prism
5 faces 9 edges
6 vertices

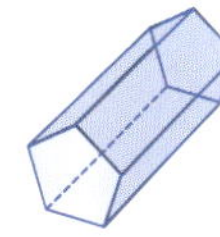

pentagonal prism
7 faces 15 edges
10 vertices

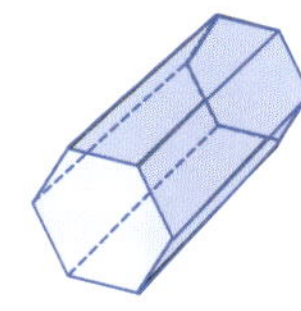

hexagonal prism
8 faces 18 edges
12 vertices

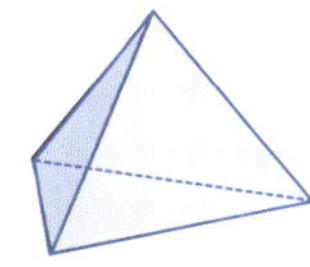

tetrahedron
(triangular pyramid)
4 faces 6 edges
4 vertices

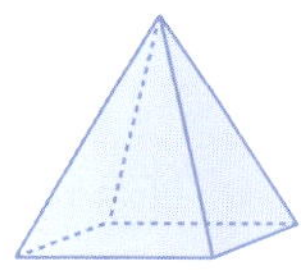

square pyramid
5 faces 8 edges
5 vertices

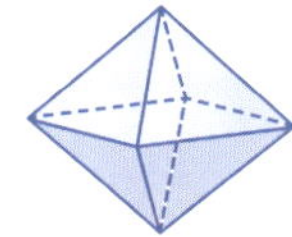

octahedron
8 faces 12 edges
6 vertices

Maths Facts

Space

Nets of 3D objects

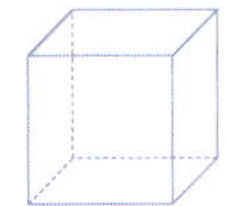
cube
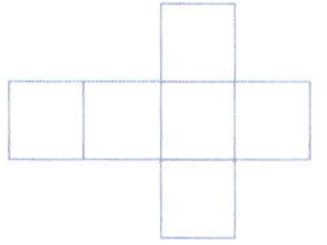

pentagonal prism

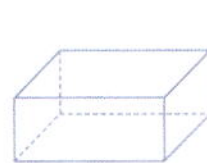
rectangular prism

hexagonal prism

cylinder
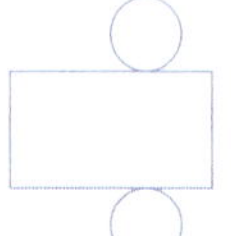

tetrahedron (triangular pyramid)
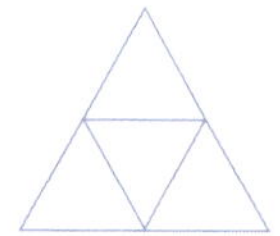

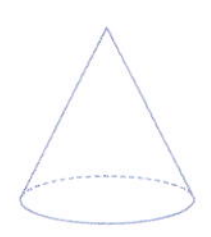
cone
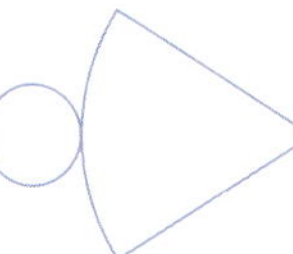

square pyramid
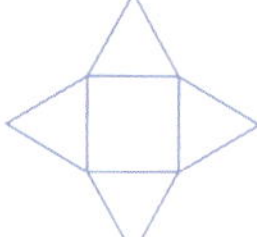

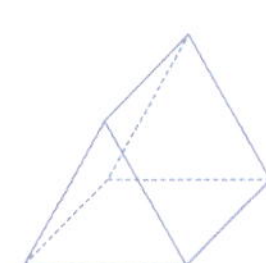
triangular prism
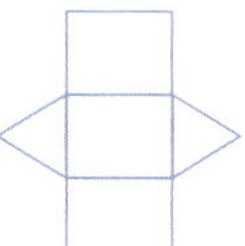

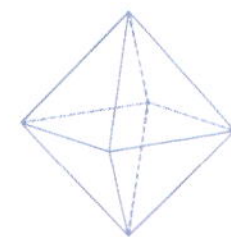
octahedron
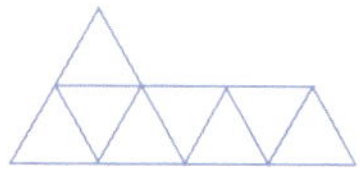

Statistics

Pictograms / Picture graphs
A graph in which data is represented by pictures. One picture could represent one unit or many.

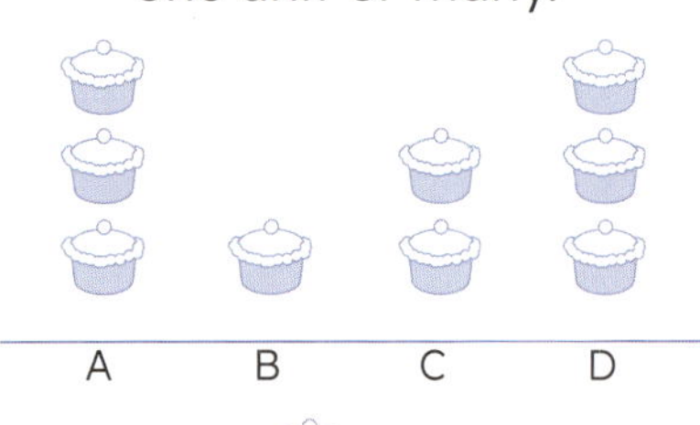

Pie charts
A graph in which sections of a circle are used to show parts of a total or whole.

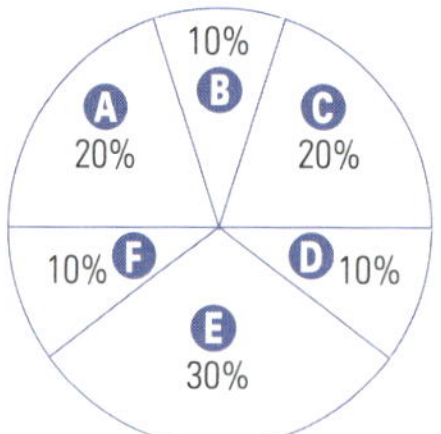

Bar graphs / Column graphs
A graph which represents information regarding frequency of outcomes using bars or columns.

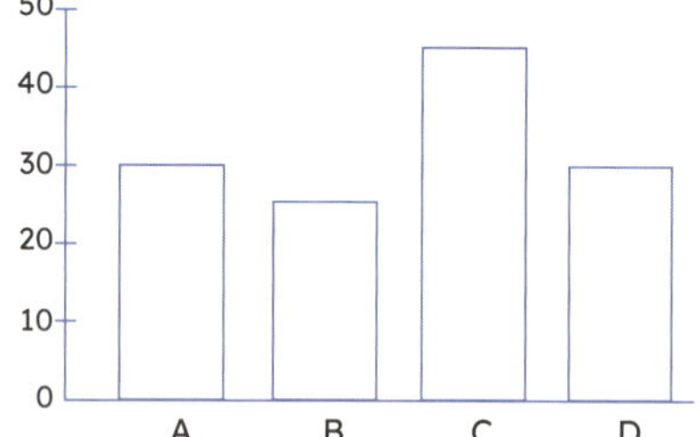

Probability

Chance

The likelihood of an event occurring.

Example
When flipping a coin, there is a 1 in 2 chance of the coin landing on heads.

1 in 2 or $\frac{1}{2}$ or 0.5 or 50%

Working Out Space

Working Out Space